INTERNATIONAL
GCSE
(9–1)

TREVOR JOHNSON
TONY CLOUGH

Mathematics
for Edexcel Specification A
Practice Book

THIRD
EDITION

HODDER
EDUCATION
AN HACHETTE UK COMPANY

Every effort has been made to trace all copyright holders, but if any have been inadvertently overlooked the publishers will be pleased to make the necessary arrangements at the first opportunity.

Although every effort has been made to ensure that website addresses are correct at time of going to press, Hodder Education cannot be held responsible for the content of any website mentioned in this book. It is sometimes possible to find a relocated web page by typing in the address of the home page for a website in the URL window of your browser.

Hachette's policy is to use papers that are natural, renewable and recyclable products and made from wood grown in sustainable forests. The logging and manufacturing processes are expected to conform to the environmental regulations of the country of origin.

Orders: please contact Bookpoint Ltd, 130 Milton Park, Abingdon, Oxon OX14 4SB. Telephone: (44) 01235 827720. Fax: (44) 01235 400454. Lines are open 9.00 – 5.00, Monday to Saturday, with a 24-hour message answering service. Visit our website at www.hoddereducation.co.uk

© Trevor Johnson, Tony Clough, 2011
First published by
Hodder Education,
an Hachette UK Company
Carmelite House, 50 Victoria Embankment
London, EC4Y 0DZ

This third edition published in 2016

Impression number   5   4
Year   2020   2019   2018

Cover illustration © Natalya Belinskaya/123.RF.com
Typeset in TimesTenLTStd-Roman, 11/13 pts. by Aptara, Inc.
Printed in Dubai

A catalogue record for this title is available from the British Library

ISBN: 978 1 471 88903 5

CONTENTS

INTRODUCTION

International GCSE Mathematics Practice for Edexcel, written by authors who were senior examiners for this examination, comprises almost 700 exam-type questions and so will be a valuable additional resource for all schools preparing candidates for Edexcel's IGCSE mathematics Higher tier papers, regardless of the text books being used in lessons.

In this third edition, the authors have taken account of the recent additions, mainly algebra, to the content of Specification A, which will be examined for the first time in 2018. There are over 100 additional questions addressing every part of the new syllabus. As it is envisaged that this book will often be used as part of students' final preparation for the exam, to provide appropriately varied practice covering all the relevant grades, each section contains a variety of questions which are not in any particular order, although questions on the new topics often appear at the ends of exercises.

From 2018, 30% of the questions in the examination will involve problem solving and 20% will test mathematical reasoning, interpretation and communication. Although, of course, many of the questions in earlier editions of this book addressed these aspects of mathematics, the authors felt it would be helpful to provide students with further practice in these skills and so have added an extra exercise of questions for this purpose.

For each question in the book, there is an indication of the mark tariff it might carry in an exam but this should be regarded only as a guide. Short numerical, algebraic and some graphical answers are also provided.

The authors would like to thank the staff of Hodder Education for the valuable help and advice they have given.

Trevor Johnson, Tony Clough

March 2016

Number

1 Decimals

1   Karl has a mobile phone. Last month, he paid 12.8p for each minute of calls plus a monthly charge of £15.25
Last month, Karl made 246 minutes of calls.
Work out the **total** amount that Karl paid last month.

(Total 3 marks)

2   Anna is a freelance typist. She types at a rate of 60 words per minute.
Anna charges 0.4p per word.
   a) Work out how much Anna charges for one hour's typing.   **(2)**

Anna types a document of 14 000 words.
   b) i) How much will Anna charge for typing the document?
   ii) How long will it take Anna to type the document?   **(5)**

(Total 7 marks)

3   Hassan buys a new car.
The car's petrol tank has a capacity of 58.5 litres.
The cost of petrol is 95.1p per litre, of which 65.7p is tax.
The car can travel 15.2 km on one litre of petrol.
   a) Work out the cost of a full tank of petrol.   **(2)**

Hassan drove 19 200 km last year.
   b) Work out how much tax Hassan paid when driving 19 200 km.
   Give your answer to the nearest pound.   **(3)**

(Total 5 marks)

4   Andre used 1521 units of electricity.
The first 207 units cost £0.1708 per unit.
The remaining units cost £0.0943 per unit.
Tax is added at 5% of the total amount.
Complete a copy of Andre's bill.

207 units at £0.1708 per unit	£ ....
...... units at £0.0943 per unit	£ ....
Total amount	£ _____
Tax at 5% of the total amount	£ ....
Amount to pay	£ _____

(Total 7 marks)

5 Work out the value of $\dfrac{17.16}{1.4 + 2.5}$

(Total 2 marks)

6 Work out the value of $\dfrac{7.3 + 9.5}{8.4 - 3.6}$

(Total 2 marks)

7 Use your calculator to work out the value of $\dfrac{8.3 - 2.7}{1.7 \times 0.4}$

Write down all the figures on your calculator display.

(Total 2 marks)

8 a) Use your calculator to work out the value of $\dfrac{8.7 \times 5.24}{8.02 + 1} - 3.6$

Write down all the figures on your calculator display. **(2)**

b) Give your answer to part **a)** correct to 3 significant figures. **(1)**

(Total 3 marks)

9 a) Use your calculator to work out the value of $2.7 - \dfrac{9.7}{1.9 + 1.7}$

Write down all the figures on your calculator display. **(2)**

b) Give your answer to part **a)** correct to 2 significant figures. **(1)**

(Total 3 marks)

10 a) Use your calculator to work out the value of $\dfrac{(12.71 - 3.9)^2}{3.7 + 5.43}$

Write down all the figures on your calculator display. **(2)**

b) Give your answer to part **a)** correct to 2 decimal places. **(1)**

(Total 3 marks)

11 Show that $0.35 + \dfrac{2}{7} = \dfrac{89}{140}$

(Total 3 marks)

12 Prove that the recurring decimal $0.\dot{7} = \dfrac{7}{9}$

(Total 2 marks)

13 Convert the recurring decimal $0.6\dot{3}$ to a fraction.
Give your answer in its simplest form.

(Total 2 marks)

14 x is an integer such that $1 \leqslant x < 10$
y is an integer such that $1 \leqslant y < 10$

Prove that $0.0\dot{x}\dot{y} = \dfrac{10x + y}{990}$

(Total 2 marks)

15 Prove that the recurring decimal $0.5\dot{3} = \dfrac{8}{15}$

(Total 2 marks)

16 Convert the recurring decimal $0.6\dot{4}$ to a fraction.
Give your answer in its simplest form.

(Total 2 marks)

17 Bhavana says:

'When the fraction $\dfrac{n}{105}$ is converted to a decimal, it never gives a terminating decimal.'

a) Find a value of n less than 105, which shows that Bhavana is wrong.

b) Write down the type of number n must be, when $\dfrac{n}{105}$ gives a terminating decimal.

(Total 2 marks)

2 Powers and roots

1 Calculate the value of $\sqrt{3.7^3 - 4.2^2}$
Write down all the figures on your calculator display.

(Total 2 marks)

2 a) Use your calculator to work out the value of $\sqrt{\dfrac{2.7}{1.2^2 + 0.9}}$

Write down all the figures on your calculator display. **(2)**

b) Give your answer to part **a)** correct to 3 significant figures. **(1)**

(Total 3 marks)

3 a) Use your calculator to work out the value of $\dfrac{\sqrt{4.3 + 2.9^3}}{1.3} + 3.6^2$

Write down all the figures on your calculator display. **(2)**

b) Give your answer to part **a)** correct to 2 decimal places. **(1)**

(Total 3 marks)

4 a) Calculate the value of $\dfrac{(4.6^2 - 1.9^3)}{\sqrt{2.4 + 4.8}}$

Write down all the figures on your calculator display. **(2)**

b) Give your answer to part **a)** correct to 4 significant figures. **(1)**

(Total 3 marks)

5 Express 175 as the product of its prime factors.

(Total 2 marks)

6 Express 392 as a product of powers of its prime factors.

(Total 3 marks)

7 Express 300 as a product of powers of its prime factors.

(Total 3 marks)

8 a) i) Express 108 as a product of powers of its prime factors.
   ii) Express 144 as a product of powers of its prime factors. **(4)**
b) Using your answers to part **a)**, or otherwise, work out the Highest Common Factor of 108 and 144. **(2)**

(Total 6 marks)

9 **a)** Express 150 as the product of its prime factors. (2)
   b) Work out the Lowest Common Multiple of 90 and 150 (2)

                                                (Total 4 marks)

10 **a)** Find the Highest Common Factor of 105 and 147 (2)
    b) Find the Lowest Common Multiple of 105 and 147 (2)

                                                (Total 4 marks)

11

Symbols
$+$ $-$ $\times$ $\div$ ( )

Using only symbols from the box, make the following into true statements.
a) 6   4   2 = 20 (1)
b) 6   4   2 = 3.5 (1)
c) 6   4   2 = $\frac{3}{4}$ (1)

                                                (Total 3 marks)

12 **a)** Find, as a fraction, the value of $\dfrac{4+2^3}{(4+2)^2}$

    b) Write your answer to part **a)** as a decimal.

                                                (Total 3 marks)

13 **a)** Simplify, leaving your answer in index form
      i) $5^4 \times 5^8$ **ii)** $9^7 \div 9^4$ (2)
    b) Solve $\dfrac{3^8 \times 3^5}{3^n} = 3^4$ (2)

                                                (Total 4 marks)

14 **a)** Simplify, leaving your answer in index form
      i) $4^7 \times 4$ **ii)** $6^3 \div 6^5$ (2)
    b) $7^y = 1$
      Find the value of y. (1)

                                                (Total 3 marks)

15 Evaluate the following.
Give your answers as fractions.

a) 3^{-4} **(1)**

b) $\sqrt[3]{\left(\dfrac{8}{27}\right)^2}$ **(1)**

c) $\left(\dfrac{\sqrt{6}}{4}\right)^4$ **(1)**

(Total 3 marks)

---

16 a) Express $27^{\frac{1}{2}}$ as a power of 3 **(2)**

b) Express $\sqrt{2}$ as a power of 8 **(2)**

c) Express $\dfrac{\sqrt{2}}{16}$ as a power of 2 **(3)**

(Total 7 marks)

---

17 Given that x and y are positive integers such that $(2+\sqrt{x})(5+\sqrt{x}) = y + 7\sqrt{3}$ find the value of x and the value of y.

(Total 3 marks)

---

18 $(3 - \sqrt{x})^2 = y - 6\sqrt{5}$ where x and y are positive integers.

Find the value of x and the value of y.

(Total 3 marks)

---

19 Given that k and n are positive integers such that $\sqrt{n}(2 + k\sqrt{n}) = 15 + 2\sqrt{3}$, find the value of k and the value of n.

(Total 3 marks)

---

20 $(6 + \sqrt{n})(6 - \sqrt{n}) = 26$ where n is a positive integer.

Find the value of n.

(Total 2 marks)

In Questions 21–34, you must show your working clearly.

21 Show that $(2 - \sqrt{3})^2 = 7 - 4\sqrt{3}$

(Total 2 marks)

22 Express $\sqrt{147}$ in the form $k\sqrt{3}$ where k is a positive integer.

(Total 2 marks)

23 Show that $\sqrt{32} - \sqrt{8} = 2\sqrt{2}$

(Total 2 marks)

24 Express $(1 + 2\sqrt{3})^2$ in the form $a + b\sqrt{3}$ where a and b are positive integers.

(Total 2 marks)

25 Given that a and b are integers such that $(\sqrt{a} - \sqrt{12a})^2 = 26 - b\sqrt{3}$ find the value of a and the value of b.

(Total 4 marks)

26 Rationalise the denominator in $\dfrac{6}{\sqrt{3}}$ and give your answer as simply as possible.

(Total 2 marks)

27 Express $\dfrac{\sqrt{3}}{\sqrt{5}}$ in the form $\dfrac{\sqrt{n}}{5}$ where n is a positive integer.

(Total 2 marks)

28 Express $\left(\dfrac{1}{\sqrt{2}}\right)^7$ in the form $\dfrac{\sqrt{2}}{n}$ where n is a positive integer.

(Total 3 marks)

29 a) Expand and simplify $(1 - \sqrt{3})^2$ **(2)**

   b) Hence show that $\dfrac{\left(1 - \sqrt{3}\right)^2}{\sqrt{3}} = -2 + \dfrac{4\sqrt{3}}{3}$ **(3)**

(Total 5 marks)

30 Rationalise the denominator in $\dfrac{1}{3 + \sqrt{5}}$

(Total 3 marks)

7

31 Rationalise the denominator in $\dfrac{\sqrt{3}}{2-\sqrt{3}}$ and give your answer in the form $a + b\sqrt{c}$ where a, b and c are positive integers.

(Total 3 marks)

32 Rationalise the denominator in $\dfrac{3+\sqrt{2}}{\sqrt{2}-1}$ and give your answer as simply as possible.

(Total 3 marks)

33 Express $\dfrac{5-2\sqrt{7}}{\sqrt{7}+2}$ in the form $a + b\sqrt{c}$ where a, b and c are integers.

(Total 4 marks)

34 Show that $\dfrac{5\sqrt{3}+2}{\sqrt{3}+1} = \dfrac{13}{2} - \dfrac{3\sqrt{3}}{2}$

(Total 4 marks)

3 Fractions

1 Show that $\frac{1}{3} + \frac{2}{5} = \frac{11}{15}$

(Total 2 marks)

2 Show that $\frac{4}{15} \times 6 = 1\frac{3}{5}$

(Total 2 marks)

3 Show that $\frac{6}{7} \times \frac{1}{9} = \frac{2}{21}$

(Total 2 marks)

4 Show that $\frac{2}{9} \div \frac{1}{4} = \frac{8}{9}$

(Total 2 marks)

5 Show that $\frac{11}{12} - \frac{2}{3} = \frac{1}{4}$

(Total 2 marks)

6 Show that $\frac{9}{10} \times \frac{2}{3} = \frac{3}{5}$

(Total 2 marks)

7 Show that $\frac{3}{8} \div \frac{5}{12} = \frac{9}{10}$

(Total 2 marks)

8 Show that $\frac{3}{4} + \frac{4}{9} = 1\frac{7}{36}$

(Total 2 marks)

9 Show that $\frac{8}{9} \div 6 = \frac{4}{27}$

(Total 2 marks)

10 Show that $3\frac{2}{9} + 1\frac{5}{6} = 5\frac{1}{18}$

(Total 3 marks)

11 Show that $2\frac{2}{5} \times 1\frac{1}{9} = 2\frac{2}{3}$

(Total 3 marks)

12 Show that $4\frac{3}{4} - 1\frac{2}{3} = 3\frac{1}{12}$

(Total 3 marks)

13 Show that $3\frac{1}{3} \div 7\frac{1}{2} = \frac{4}{9}$

(Total 3 marks)

14 Show that $4\frac{1}{6} - 2\frac{4}{5} = 1\frac{11}{30}$

(Total 3 marks)

15 Show that $6\frac{2}{3} \div 1\frac{7}{8} = 3\frac{5}{9}$

(Total 3 marks)

16 Last season, Hodder Hurricanes won $\frac{5}{12}$ of its matches, drew $\frac{2}{9}$ and lost the rest.
Show that it lost $\frac{13}{36}$ of its matches.

(Total 3 marks)

17 Of the teachers in a school, $\frac{3}{4}$ are women.
Of the women teachers in the school, $\frac{5}{6}$ are married.
Work out the fraction of teachers in the school who are married women.
Give your fraction in its simplest form.

(Total 2 marks)

18 Pat's dog eats $\frac{2}{3}$ of a tin of food each day.
She has 8 tins of dog food.
How many days will the 8 tins last?

(Total 2 marks)

19 Chloe, Daisy and Emily share some money.
Chloe receives $\frac{5}{8}$ of the money.
Daisy receives $\frac{1}{6}$ of the money.
Show that Emily receives $\frac{5}{24}$ of the money.

(Total 3 marks)

20 $\frac{7}{12}$ of the passengers on a coach are boys.
The rest of the passengers are girls.
35 of the passengers are boys.
How many of the passengers are girls?

(Total 3 marks)

21 $\frac{3}{4}$ of Joshua's books are paperbacks.

$\frac{4}{9}$ of his paperback books are science fiction.

What fraction of his books are science fiction paperbacks?
Give your fraction in its simplest form.

(Total 2 marks)

22 $\frac{8}{9}$ of an iceberg lies below the water surface.

Calculate the total volume of the iceberg, if 13 200 m^3 of the iceberg lies below the water surface.

(Total 2 marks)

23 $\frac{5}{9}$ of the students in a school are girls.

$\frac{3}{5}$ of these girls own a bicycle.

$\frac{3}{4}$ of the boys in the school own a bicycle.

What fraction of the students in the school own a bicycle?
Give your fraction in its simplest form.

(Total 3 marks)

24 A bag contains only red beads, white beads and blue beads.

$\frac{5}{12}$ of the beads are red.

$\frac{3}{8}$ of the beads are white.

Work out the smallest possible number of beads that the bag could contain.

(Total 2 marks)

25 Lian bought 450 shirts to sell in her shop.
She paid $10 for each shirt.

In May, she sold $\frac{7}{9}$ of the 450 shirts.
She sold these shirts at $18 each.

In June, she reduced the price of each shirt by $\frac{1}{3}$
She sold the remaining shirts at this price.

a) Calculate the total profit that Lian made on the shirts. **(4)**

b) Express her total profit as a fraction of the amount she paid for the shirts.
Give your fraction in its simplest form. **(1)**

(Total 5 marks)

4 Percentages

1 Express 63 as a percentage of 350

(Total 2 marks)

2 Express 42 seconds as a percentage of 5 minutes.

(Total 3 marks)

3 Benita's weekly pay was increased from £450 to £468
Work out the percentage increase in Benita's weekly pay.

(Total 3 marks)

4 Paulo invested $7500 at an interest rate of 5.4% per year.
Work out the total amount in his account after one year.

(Total 3 marks)

5 In a sale, normal prices are reduced by 12%.
  a) The normal price of a case is £45
    Work out the sale price of the case. (3)
  b) The price of a computer is reduced by £69
    Work out the normal price of the computer. (3)

(Total 6 marks)

6 Salim invested an amount of money at 6% interest.
After one year, interest was added to his account.
The amount in his account was then £901
Work out the amount of money Salim invested.

(Total 3 marks)

7 Of the number of people a company employs, 64% are men.
The company employs 112 men.
Work out the number of people the company employs.

(Total 3 marks)

8 An art dealer bought a painting for £2500 and sold it for £3300
Work out her percentage profit.

(Total 3 marks)

9   Magda got 54 out of 75 in a science test.
    a) Work out 54 out of 75 as a percentage. **(2)**

    Dwayne got 45% of the total marks in a history test.
    Dwayne got 63 marks.
    b) Work out the total number of marks for the history test. **(3)**

    (Total 5 marks)

10 A flight normally costs £650
    It is reduced by 18%.
    How much will the flight now cost?

    (Total 3 marks)

11 Jack bought a computer for $875 and later sold it for $735
    Work out his percentage loss.

    (Total 3 marks)

12 In a sale, normal prices are reduced by 35%.
    The sale price of a clock is £78
    Work out the normal price of the clock.

    (Total 3 marks)

13 During 2006, the population of Kenya increased from 34 708 000 to 36 914 000
    Calculate the percentage increase in Kenya's population.
    Give your answer correct to 1 decimal place.

    (Total 3 marks)

14 Linton's salary is £28 000
    His salary is increased by 4.5%.
    Work out Linton's new salary.

    (Total 3 marks)

15 Before dieting, Kate's weight was 75 kg.
    After dieting, her weight was 66 kg.
    Work out Kate's percentage weight loss.

    (Total 3 marks)

16 David and Gwen bought a house.
    After one year, its value had increased by 9% to £196 200
    Work out the value of the house when they bought it.

    (Total 3 marks)

17 70.9% of the Earth's surface is covered by water.
The area of the Earth's surface covered by water is 362 000 000 km².
Calculate the total surface area of the Earth.
Give your answer correct to 2 significant figures.

(Total 3 marks)

18 100 g of Bran Flakes contains 10.8 g of protein.
The Recommended Daily Amount (RDA) of protein is 90 g.
a) What percentage of the RDA of protein is provided by 100 g of Bran Flakes? **(2)**

100 g of Bran Flakes contains 16.1 mg of iron.
The Recommended Daily Amount (RDA) of iron is 14 mg.
b) What percentage of the RDA of iron is provided by 100 g of Bran Flakes? **(2)**

100 g of Bran Flakes contains 13.6 g of fibre.
13.6 g is 56.7% of the RDA of fibre.
c) Calculate the RDA of fibre.
Give your answer correct to 2 significant figures. **(3)**

(Total 7 marks)

19 The number of candidates who took maths IGCSE was 88.9% greater in May 2007 than the number of candidates in May 2006
In May 2007, 18 800 candidates took maths IGCSE.
Calculate the number of candidates who took maths IGCSE in May 2006

(Total 3 marks)

20 *Solarhols* make a cancellation charge if a customer cancels a holiday less than six weeks before departure.
The cancellation charge is a percentage of the cost of the holiday.
Hassan booked a holiday with *Solarhols*.
He cancelled his holiday three weeks before the departure date.
He had to pay a cancellation charge of 40% of the cost of the holiday.
His cancellation charge was £408
Work out the cost of his holiday.

(Total 3 marks)

21 Tamsin bought a car.
The cost of the car was $8640
She paid a deposit of 35% of the cost of the car.
She paid the remainder of the cost in 36 equal monthly instalments.
Work out the monthly instalment Tamsin paid.

(Total 4 marks)

22 Rohan opened an account with £800 at the London Bank.
After one year, the bank paid him interest.
He then had £836 in his account.
Work out, as a percentage, London Bank's interest rate.

(Total 3 marks)

23 The length of a new curtain was 85 cm.
When it was washed, the curtain shrank by 4%.
Work out the length of the curtain after it had been washed.

(Total 3 marks)

24 Usave Motor Insurance reduces the premiums its customers pay for car insurance
by a certain percentage, according to the number of consecutive years for which the
customer has not made a claim.
This percentage reduction in premiums is called the 'No-claims bonus'.
The table shows Usave's No-claims bonuses.

Number of consecutive years without a claim	No-claims bonus
1	35%
2	45%
3	55%
4	60%
5 or more	65%

Leela has not made a claim during the last 2 years.
Without the No-claims bonus, her premium would be £360
a) Calculate her premium with the No-claims bonus. **(3)**

Dean has not made a claim during the last 7 years.
With the No-claims bonus, his premium is £182
b) Calculate his premium without the No-claims bonus. **(3)**

(Total 6 marks)

25 Sachin invested £6000 for 2 years at 5% per annum compound interest.
Calculate the value of his investment at the end of 2 years.

(Total 3 marks)

26 $5000 is invested for 3 years at 4% per annum compound interest.
Work out the total interest earned over the 3 years.

(Total 3 marks)

27 Liz bought a car for £12 500
The car depreciated by 10% each year.
Work out the value of the car 4 years after she bought it.

(Total 3 marks)

28 Rajiv invested $20 000 at 7% per annum compound interest.
After n years, the value of his investment had grown to $30 014.61
Find the value of n.

(Total 3 marks)

29 Emma invested an amount of money.
Compound interest was paid at a rate of 4.6% per annum.
After 2 years, the value of her investment was £30 088.19
Calculate the amount she invested.

(Total 3 marks)

30 Ben invested an amount of money at 5.2% per annum compound interest.
Calculate the number of years after which the value of his investment had doubled.

(Total 3 marks)

31 An increase of 30% is followed by an increase of 40%.
Find the total percentage increase.

(Total 3 marks)

32 The value of a new car decreases by 20% in the first year and by 15% in the second year.
Calculate the total percentage decrease in its value in the two years.

(Total 3 marks)

33 Leela says that a decrease of 35% followed by an increase of 60% is the same as an increase of 25%.
Is Leela correct? You must show working to justify your answer.

(Total 3 marks)

34 When *Easy PC* sells computers at their normal prices, it makes a profit of 30%.
In a sale, normal prices are reduced by 10%.
Calculate the percentage profit on a computer sold in the sale.

(Total 3 marks)

35 Find the total percentage change when an increase of 60% is followed by a decrease of 50% and then by a decrease of 10%.

(Total 4 marks)

5 Ratio and proportion

1   The recommended daily amount of fibre for men is 24 grams.
    The recommended daily amount of carbohydrates for men is 300 grams.
    Find the ratio of the recommended daily amount of carbohydrates for men to the
    recommended daily amount of fibre for men.
    Give your ratio in the form $1 : n$

                                                                    (Total 2 marks)

2   The population of Malta is 4×10^5
    The population of the United Kingdom is 6×10^7
    Find the ratio of the population of Malta to the population of the United Kingdom.
    Give your ratio in the form $1 : n$

                                                                    (Total 2 marks)

3   The length of an aeroplane is 7.5 m.
    A scale model is made of the aeroplane.
    The length of the scale model is 30 cm.
    a) Express the scale of the model in the form $1 : n$ **(3)**

    The wingspan of the scale model is 36 cm.
    b) Work out the wingspan of the real aeroplane.
        Give your answer in metres.   **(3)**

                                                                    (Total 6 marks)

4   A total of 64 320 people watched the football match between United and City.
    The ratio of United supporters to City supporters was 15 : 1
    Work out the number of United supporters who watched the match.

                                                                    (Total 2 marks)

5   John and Pavinder share some money in the ratio 5 : 7
    John receives $240
    Work out the amount of money that Pavinder receives.

                                                                    (Total 2 marks)

6   Brass is made from copper and zinc in the ratio 13 : 7 by weight.
    a) Work out the weight of copper and the weight of zinc in 5 kg of brass.   **(3)**

    A brass ornament contains 350 grams of zinc.
    b) Work out the weight of copper in the ornament.   **(2)**

                                                                    (Total 5 marks)

7 The lengths of the sides of an isosceles triangle are in the ratios 3 : 3 : 5
The length of the equal sides of the triangle is 6.6 cm.
Work out the perimeter of the triangle.

(Total 3 marks)

8 The sizes of the angles of a triangle are in the ratios 2 : 3 : 7
Work out the size of the obtuse angle of the triangle.

(Total 3 marks)

9 Rajesh runs a distance of $2\frac{2}{3}$ km each day.
How far, in kilometres, does Rajesh run in 7 days?
Give your answer as a mixed number.

(Total 2 marks)

10 The amount of money Zhanna is paid is directly proportional to the number of
computers she sells. Last month Zhanna sold 28 computers and was paid $1512
   a) Work out the number of computers Zhanna sold when she was paid $2106   **(2)**
   b) Work out the amount of money Zhanna is paid for selling 21 computers.   **(2)**

(Total 4 marks)

11 The scale of a map is 1 : 5 000 000
On the map, the distance between two airports is 3.6 cm.
   a) Work out the real distance between the airports.
      Give your answer in kilometres.   **(3)**

The real distance between two railway stations is 30 km.
   b) Work out the distance on the map between the two railway stations.   **(2)**

(Total 5 marks)

12 Here are the ingredients needed to make a pie for 4 people.

<u>Ingredients for 4 people</u>
150 g of pastry
450 g of potatoes
220 g of onions
320 g of bacon

Work out the amount of each ingredient needed to make a pie for 10 people.

(Total 3 marks)

13 Kurt went on holiday to England.
He changed €1800 to pounds.
The exchange rate was €1 = £0.69

a) Work out the number of pounds Kurt received. **(2)**

Kurt returned home with £198
He changed his £198 to euros.
The new exchange rate was €1 = £0.66

b) Work out the number of euros Kurt received. **(2)**

(Total 4 marks)

6 Standard form

1 **a)** Write the number 230 000 in standard form. **(1)**
   b) Write 6.3×10^{-3} as an ordinary number. **(1)**

   (Total 2 marks)

2 **a)** Write the number 0.000 47 in standard form. **(1)**
   b) Write 3.7×10^6 as an ordinary number. **(1)**
   c) Work out the value of $\dfrac{5 \times 10^3}{8 \times 10^{-5}}$

   Give your answer in standard form. **(1)**

   (Total 3 marks)

3 The distance between Mercury and the Sun is 58 000 000 km.
   a) Write the number 58 000 000 in standard form. **(1)**

   Uranus is 50 times as far from the Sun as Mercury.
   b) Calculate the distance between Uranus and the Sun.
   Give your answer in standard form. **(2)**

   (Total 3 marks)

4 Work out the value of $(6 \times 10^5) \times (8 \times 10^{-4})$.
   Give your answer in standard form.

   (Total 2 marks)

5 Work out the value of $(7.2 \times 10^4) + (4.7 \times 10^3)$.
   Give your answer in standard form.

   (Total 2 marks)

6 The age of the Universe is approximately 15 000 million years.
   a) Write the number 15 000 million in standard form. **(1)**

   1 gigayear $= 10^9$ years.
   b) Express the age of the Universe in gigayears. **(2)**

   (Total 3 marks)

7 The mass of a proton is 0.000 000 000 000 000 000 000 001 67 grams.
  a) Write the number 0.000 000 000 000 000 000 000 001 67 in standard form. **(1)**

  The mass of a proton is 1850 times greater than the mass of an electron.
  b) Calculate the mass, in grams, of an electron.
    Give your answer in standard form correct to 2 significant figures. **(2)**

  (Total 3 marks)

8 $P = 2a + 5b$
  $a = 3.5 \times 10^6$ and $b = 2.7 \times 10^7$
  Work out the value of P.
  Give your answer in standard form.

  (Total 2 marks)

9 A spherical planet has a radius of 6800 km.
  Calculate the surface area, in km², of the planet.
  Give your answer in standard form correct to 2 significant figures.

  (Total 3 marks)

10 $y = \dfrac{a+b}{ab}$
  $a = 2 \times 10^4$ and $b = 8 \times 10^5$
  Work out the value of y.
  Give your answer in standard form correct to 3 significant figures.

  (Total 3 marks)

11 A supercomputer can perform 12 300 000 000 000 calculations per second.
  a) Write the number 12 300 000 000 000 in standard form. **(1)**
  b) Work out the number of calculations this computer can perform in 4 hours.
    Give your answer in standard form correct to 2 significant figures. **(2)**

  1 nanosecond = 10^{-9} seconds.
  c) Work out the time, in nanoseconds, this computer takes to perform one
    calculation.
    Give your answer in standard form correct to 2 significant figures. **(2)**

  (Total 5 marks)

12 $F = 3p - 4q$
  $p = 2 \times 10^{-5}$ and $q = 8 \times 10^{-6}$
  Work out the value of F.
  Give your answer in standard form.

  (Total 2 marks)

13 The Andromeda Galaxy is the most distant galaxy that can be seen with the naked eye from Earth.

The distance from the Andromeda Galaxy to Earth is 2.19×10^{19} km.

Light travels 9.46×10^{12} km in a year.

Calculate the time, in years, light takes to travel from the Andromeda Galaxy to Earth.

Give your answer in standard form correct to 2 significant figures.

(Total 3 marks)

14 $T = \dfrac{xy}{x-y}$

$x = 5.3 \times 10^{-3}$ and $y = 6.7 \times 10^{-4}$

Work out the value of T.

Give your answer in standard form correct to 3 significant figures.

(Total 3 marks)

15 1 litre of paint covers an area of 12 m².

Work out the thickness, in centimetres, of the coat of paint.

Give your answer in standard form correct to 2 significant figures.

(Total 4 marks)

16 $p^2 = \dfrac{2a}{b}$

$a = 7.3 \times 10^{13}$ and $b = 2.9 \times 10^{-4}$

Work out the value of p.

Give your answer in standard form correct to 2 significant figures.

(Total 3 marks)

17 The mass of an atom can be expressed in atomic mass units.

1 atomic mass unit (amu) = 1.66×10^{-27} kg.

a) The weight of 1 atom of carbon 12 is 12 amu.

Express 12 amu in kg.

Give your answer in standard form. **(2)**

b) The weight of 1 atom of gold is 3.27×10^{-25} kg.

Express 3.27×10^{-25} kg in amu.

Give your answer correct to the nearest whole number. **(2)**

(Total 4 marks)

18 1 light year $= 9.46 \times 10^{12}$ km.
  a) The distance from the Earth to the star Alpha Centauri is 4.35 light years.
       Work out the distance, in kilometres, from the Earth to the star Alpha Centauri.
       Give your answer in standard form correct to 2 significant figures.   **(2)**
  b) The distance from the Earth to the centre of the Milky Way is 2.65×10^{17} km.
       Express 2.65×10^{17} km in light years.
       Give your answer in standard form correct to 2 significant figures.   **(2)**

(Total 4 marks)

19 Muons and taus are subatomic particles.
     The lifetime of a muon is 2.2×10^{-6} seconds.
  a) Write 2.2×10^{-6} as an ordinary number.   **(1)**

     The lifetime of a muon is 7.56×10^{6} times greater than the lifetime of a tau.
  b) Calculate the lifetime, in seconds, of a tau.
       Give your answer in standard form correct to 3 significant figures.   **(2)**

(Total 3 marks)

20 The mass of the Earth is 5.97×10^{21} tonnes.
     The mass of the Moon is 7.36×10^{19} tonnes.
  a) How many times greater is the mass of the Earth than the mass of the Moon?
       Give your answer correct to the nearest whole number.   **(2)**

     The mass of Mars is 11% of the mass of the Earth.
  b) Work out the difference in mass, in tonnes, between Mars and the Moon.
       Give your answer in standard form correct to 2 significant figures.   **(2)**

(Total 4 marks)

21 $x = 2 \times 10^{m}$ and $y = 3 \times 10^{n}$ where m and n are integers.
     Find an expression, in standard form, for xy.

(Total 2 marks)

22 $x = 6 \times 10^{m}$ and $y = 4 \times 10^{n}$ where m and n are integers.
     Find an expression, in standard form, for:

  a) xy **(2)**
  b) $\dfrac{x}{y}$ **(2)**

(Total 4 marks)

23 $x = 7 \times 10^{n}$ where n is an integer.
     Find an expression, in standard form, for x^2
     Give your expression as simply as possible.

(Total 3 marks)

24 $x = 5 \times 10^m$ and $y = 8 \times 10^n$ where m and n are integers.

Find an expression, in standard form, for $\dfrac{x}{y}$.

(Total 3 marks)

25 $3.2 \times 10^8 + a \times 10^7 = 3.8 \times 10^8$
Find the value of a.

(Total 2 marks)

26 $(7 \times 10^5) \times (a \times 10^n) = 6.3 \times 10^{-4}$ where $1 \leqslant a < 10$ and n is an integer.
Find the value of a and the value of n.

(Total 3 marks)

27 $\dfrac{7 \times 10^6}{a \times 10^n} = 8.75 \times 10^9$ where $1 \leqslant a < 10$ and n is an integer.

Find the value of a and the value of n.

(Total 3 marks)

28 $x = 4 \times 10^m$ and $y = 2 \times 10^n$ where m and n are integers.

$xy = 8 \times 10^2$ and $\dfrac{x}{y} = 2 \times 10^8$

Find
a) the value of m **(2)**
b) the value of n. **(2)**

(Total 4 marks)

7 Degree of accuracy

1 Sophie wants to work out an estimate for the value of $2.8 + 34.1 \times 0.37$ without using a calculator.
Write down, for each number, an approximate value she could use.

(Total 2 marks)

2 Dilip wants to work out an estimate for the value of $\dfrac{42.7 \times 9.3}{27.2}$ without using a calculator.
a) Round each number in Dilip's calculation to 1 significant figure. **(2)**
b) Use your rounded numbers to work out an estimate for the value
of $\dfrac{42.7 \times 9.3}{27.2}$ **(1)**
c) Without using a calculator, explain why your answer to part **b)** should be smaller than the exact answer. **(1)**

(Total 4 marks)

3 Alice wants to work out an estimate for the value of $\dfrac{36.8 \times 78.2}{0.53 \times 4.2}$ without using a calculator.
a) Round each number in Alice's calculation to 1 significant figure. **(2)**
b) Use your rounded numbers to work out an estimate for the value
of $\dfrac{36.8 \times 78.2}{0.53 \times 4.2}$
Give your answer correct to 1 significant figure. **(2)**
c) Without using a calculator, explain why your answer to part **b)** should be larger than the exact answer. **(2)**

(Total 6 marks)

4 The weight of a cat is 6.4 kg, correct to 2 significant figures.
a) Write down the upper bound for the weight of the cat.
b) Write down the lower bound for the weight of the cat.

(Total 2 marks)

5 The diameter of a CD is 11.8 cm, correct to 3 significant figures.
a) Write down the upper bound for the diameter of a CD.
b) Write down the lower bound for the diameter of a CD.

(Total 2 marks)

6 The weight of a book is 0.57 kg, correct to 2 decimal places.
Calculate the lower bound for the weight of 40 of these books.

(Total 2 marks)

7 The length of a side of a square is 7.4 cm, correct to 1 decimal place.
   a) Work out
      i) the upper bound for the perimeter of the square
      ii) the lower bound for the perimeter of the square. **(3)**
   b) **i)** Give the perimeter of the square to an appropriate degree of accuracy.
      ii) Explain how you obtained your answer. **(2)**

(Total 5 marks)

8 The distance from London to Birmingham is 179 km, correct to the nearest kilometre.
On Monday, Sean drove from London to Birmingham and back.
He repeated the journey on Tuesday and on Wednesday.
Calculate the upper bound for the total distance Sean drove.

(Total 2 marks)

9 Correct to 1 significant figure, $x = 9$ and $y = 6$
   a) Calculate the lower bound for the value of $x - y$. **(2)**
   b) Calculate the upper bound for the value of xy. **(2)**

(Total 4 marks)

10 The perimeter of a regular pentagon is 24 cm, correct to 2 significant figures.
Calculate the lower bound for the length of each side.

(Total 2 marks)

11 Correct to the nearest centimetre, Alan's height is 1.76 m and Nina's height is 1.64 m.
Calculate the upper bound for the difference between their heights.

(Total 3 marks)

12 Correct to 1 significant figure, $x = 8$ and $y = 2$

Calculate the lower bound for the value of $\dfrac{x}{y}$.

(Total 3 marks)

13 The circumference of a circle is 23.4 cm, correct to 3 significant figures.
   a) Calculate the upper bound for the diameter of the circle.
      Write down all the figures on your calculator display. **(2)**
   b) Give the diameter of the circle to an appropriate degree of accuracy.
      You must show working to explain how you obtained your answer. **(2)**

(Total 4 marks)

14 Correct to 2 significant figures, the area of a rectangle is 29 cm².
Correct to 2 significant figures, the width of the rectangle is 4.8 cm.
Calculate the upper bound for the length of the rectangle.
Write down all the figures on your calculator display.

(Total 3 marks)

15 When full, a car's petrol tank holds 54 litres, correct to the nearest litre.
Correct to 1 decimal place, the car travels 15.8 km on one litre of petrol.
Calculate the lower bound for the distance the car will travel on a full tank of petrol.

(Total 3 marks)

16 Sinead ran 68 metres, correct to the nearest metre.
Her time for the run was 9.8 seconds, correct to the nearest tenth of a second.
Calculate the upper bound for her average speed in m/s.

(Total 3 marks)

17 A crane can lift a load of 3210 kg, correct to 3 significant figures.
The weight of a box is 92 kg, correct to the nearest kilogram.
Calculate the greatest number of these boxes the crane can lift.

(Total 3 marks)

18

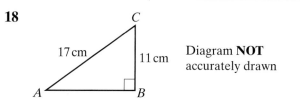

Diagram **NOT**
accurately drawn

In triangle ABC, angle $ABC = 90°$.
$BC = 11$ cm, correct to the nearest centimetre.
$AC = 17$ cm, correct to the nearest centimetre.
Calculate the upper bound for the length of AB.

(Total 4 marks)

19 The formula $I = \dfrac{W}{h^2}$ gives the Body Mass Index, I, of a person with weight
W kilograms and height h metres.
Jose's weight is 73 kg, correct to the nearest kilogram.
His height is 1.79 m, correct to the nearest centimetre.
Calculate Jose's Body Mass Index to an appropriate degree of accuracy.
You must show working to explain how you obtained your answer.

(Total 5 marks)

20

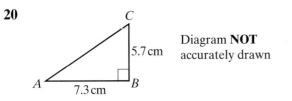

Diagram **NOT** accurately drawn

Triangle ABC is right-angled at B.
$AB = 7.3$ cm, correct to 2 significant figures.
$BC = 5.7$ cm, correct to 2 significant figures.
The line AB is horizontal.
Calculate the gradient of the line AC to an appropriate degree of accuracy.
You must show working to explain how you obtained your answer.

(Total 5 marks)

8 Set language and notation

1 $A = \{1, 2, 3, 4, 6\}$
    $B = \{1, 3, 5, 8\}$
   a) List the members of the set:
       i) $A \cup B$ **ii)** $A \cap B$.   **(2)**
   b) Explain clearly the meaning of $5 \notin A$.   **(1)**

                                                                   (Total 3 marks)

2   **a)** $A = \{$Quadrilaterals whose diagonals cross at right angles$\}$
        $B = \{$Cyclic quadrilaterals$\}$
        Write down the mathematical name for the quadrilaterals in:
        i) A **ii)** $A \cap B$.   **(2)**
   b) $\mathscr{E} = \{$Positive whole numbers less than 20$\}$
        $P = \{$Multiples of 2$\}$
        $Q = \{$Multiples of 7$\}$
        i)   Explain why it is not true that $21 \in Q$.
        ii) Is it true that $P \cap Q = \varnothing$? Explain your answer.   **(3)**

                                                                   (Total 5 marks)

3 $\mathscr{E} = \{1, 2, 3, 4, 5, 6, 7, 8, 9, 10\}$
    $P = \{1, 2, 5, 7, 9, 10\}$
   a) List the members of P'.   **(1)**

   The set Q satisfies both $Q \subset P$ **and** $n(Q) = 4$
   b) List the members of **one** set Q which satisfies both these conditions.   **(2)**

                                                                   (Total 3 marks)

4 $\mathscr{E} = \{$positive integers less than 16$\}$
    $A = \{x : x \geqslant 9\}$
    $B = \{x : x < 14\}$
   a) List the members of the set:
       i) $A \cap B$ **ii)** B'.   **(2)**
   b) What is $(A \cup B)'$?   **(1)**

                                                                   (Total 3 marks)

5 $P = \{$Prime numbers between 20 and 30$\}$
    $M = \{$Multiples of 7 between 20 and 30$\}$
   a) List the members of $P \cup M$.   **(2)**
   b) What is $M \cap P$?   **(1)**
   c) Is it true that $27 \in P$? Explain your answer.   **(1)**

                                                                   (Total 4 marks)

6 $\mathscr{E}$ = {odd numbers less than 30}
P = {factors of 36}
Q = {factors of 60}
a) **i)** Explain why $12 \notin P \cap Q$.
ii) List the members of Q. (2)
b) List the members of $P \cap Q$. (2)

(Total 4 marks)

7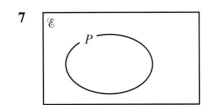

Set P is shown on the Venn diagram.
Two sets, Q and R, are such that:

$P \subset R$
$P \cap R \cap Q = \varnothing$
$R \cap Q = Q$

Complete a copy of the Venn diagram to show set Q and set R.

(Total 3 marks)

8

Statements				
$A = B$	$B \subset A$	$A' \cup B = \mathscr{E}$	$A \cap B' = A$	$A \cap B = B$

Choose a statement from the box that describes the relationship between sets A and B.

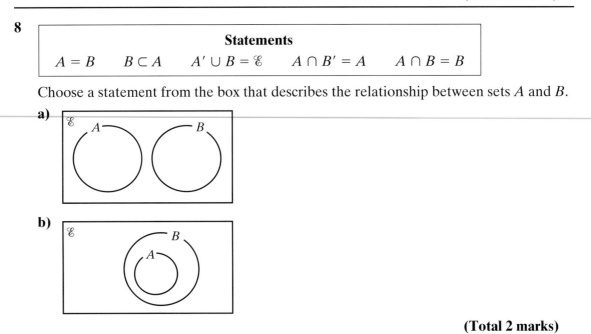

a)

b)

(Total 2 marks)

9 $\mathscr{E}$ = {positive whole numbers less than 100}
A = {multiples of 5}
B = {multiples of 2}
Sets A and B are represented by the circles in the Venn diagram:

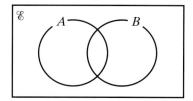

a) i)   Copy the diagram and shade the region that represents the set $(A \cup B)'$.
 ii) Write down two members of the set $(A \cup B)'$.
 iii) Write down two members of the set $A \cap B'$.   **(3)**
C = {multiples of 6}
b) i)   Copy the diagram and draw a circle, C, to represent the set C.
 ii) Find n$(A \cap B \cap C)$.   **(2)**

(Total 5 marks)

10

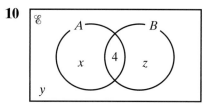

In the Venn diagram, 4, x, y and z represent the **numbers** of elements.
n$(\mathscr{E})$ = 30   n(A) = 9   n(B') = 20
a) Find the value of:
 i) x ii) y iii) z **(3)**
b) Find
 i) n$(A' \cup B)$ ii) n$((A \cup B)')$ iii) n$(A \cap B \cap B')$ **(3)**

(Total 6 marks)

11 Each student in a Year 11 group studies at least one of French, German and Spanish.

15 students study German only.
25 students study Spanish only.
20 students study French only.
6 students study German and Spanish but not French.
13 students study French and Spanish.
10 students study French and German.
x students study all three languages.

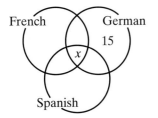

a) Write down an expression, in terms of x, for:
   i) the number of students who study French and Spanish but not German
   ii) the number of students who study French and German but not Spanish. **(2)**

There are 81 students in the group.
b) Find the value of x. **(3)**

(Total 5 marks)

---

12 There are 32 students in a class.

18 students have a brother.
10 students have both a brother and a sister.
9 students have neither a brother nor a sister.

Find the number of students in the class who have a sister.

(Total 4 marks)

Algebra

9 Algebraic manipulation

1 **a)** Simplify $a + 4b - 6a + 2b - 1$ **(2)**
    b) Multiply out $3(2p + 7)$ **(1)**
    c) Factorise $9x + 12$ **(1)**

(Total 4 marks)

2 **a)** Expand $x(3x + 2)$ **(2)**
    b) Expand and simplify $4(3x - 1) + 3(x + 5)$ **(2)**
    c) Expand and simplify $(y + 4)(y + 2)$ **(2)**

(Total 6 marks)

3 **a)** Multiply out and simplify $3(2x + 1) + 2(x + 4)$ **(2)**
    b) Multiply out $x(5x + 4y)$ **(2)**
    c) Factorise $12y - 8$ **(1)**

(Total 5 marks)

4 **a)** Expand and simplify $2(2y + 1) + 5(y - 3)$ **(2)**
    b) Factorise $15 - 20p$ **(1)**
    c) Expand and simplify $(x - 6)(x + 1)$ **(2)**

(Total 5 marks)

5 **a)** Expand $q(7q^2 + 5)$ **(2)**
    b) Expand and simplify $5(y + 1) - 4(2y + 3)$ **(2)**
    c) Factorise $x^2 - 6x$ **(2)**

(Total 6 marks)

6 **a)** Expand and simplify $7(y - 1) - 6(2y - 3)$ **(2)**
    b) Expand $2x(x - 3y)$ **(2)**

(Total 4 marks)

7 **a)** Expand and simplify $3(4x - 5) + 2(x - 3)$ **(2)**
    b) Expand $3p^3(p - 1)$ **(2)**
    c) Expand and simplify $(y - 1)(y - 5)$ **(2)**

(Total 6 marks)

8 Factorise completely $x^4y^2 - x^2y^3$

(Total 2 marks)

9 Factorise fully $20\,e^3f^2 + 25\,ef^2$

(Total 2 marks)

10 Factorise completely $27\,c^2d^4 - 18\,c^3d$

(Total 2 marks)

11 a) Factorise completely $18x - 8x^2$ **(2)**
   b) Factorise $x^2 + 9x + 20$ **(2)**

(Total 4 marks)

12 a) Expand and simplify $(3p - 2q)(2p + 5q)$ **(2)**
   b) Factorise $x^2 - 16$ **(2)**

(Total 4 marks)

13 Factorise $x^2 - 8x - 20$

(Total 2 marks)

14 Factorise $2x^2 + 9x - 5$

(Total 2 marks)

15 Factorise completely $8x^2 + 10x - 12$

(Total 3 marks)

16 Factorise completely $12x^2 - 75$

(Total 2 marks)

17 Simplify $\dfrac{2(x+2)^2}{8(x+2)}$

(Total 2 marks)

18 Simplify fully $\dfrac{x^2 + 2x - 3}{x^2 - 1}$

(Total 3 marks)

19 Simplify fully $\dfrac{x^2 - 9}{x^2 - 9x - 36}$

(Total 3 marks)

20 Simplify fully $\dfrac{x^2 + 5x + 6}{x^2 + x - 6}$

(Total 3 marks)

21 Express the algebraic fraction $\dfrac{2x^2 + 11x + 12}{4x^2 - 9}$ as simply as possible.

(Total 3 marks)

22 Simplify fully $\dfrac{2}{x - 2} + \dfrac{3x - 4}{x^2 - 5x + 6}$

(Total 6 marks)

23 Simplify fully $\dfrac{3}{x - 3} - \dfrac{x + 18}{x^2 + x - 12}$

(Total 6 marks)

24 a) Factorise $9x^2 - 1$ **(1)**
   b) Hence express as a product of its prime factors:
   i) 899
   ii) 8.99×10^4 **(4)**

(Total 5 marks)

25 Show that $1 - (x + 2) \div \left(\dfrac{x^2 - 4x - 12}{x - 8} \right)$ can be written as $\dfrac{a}{x + b}$ where a and b are integers.

(Total 4 marks)

26 Write $3 - (x + 3) \div \left(\dfrac{x^2 - 9}{x - 9} \right)$ as a single fraction.
   Simplify your answer fully.

(Total 4 marks)

27 Expand and simplify $x(x - 2)(x + 2)$

(Total 4 marks)

28 Expand and simplify $(x + 1)(x + 2)(x + 3)$

(Total 3 marks)

29 Expand and simplify $(x-3)(x-1)(x-2)$

(Total 3 marks)

30 Expand and simplify $(x-4)(x+1)(2x-1)$

(Total 3 marks)

31 Expand and simplify $(2-x)(x+3)(x-3)$

(Total 3 marks)

32 Expand and simplify $(1-x)(4-x)(x+1)$

(Total 3 marks)

33 Expand and simplify $(3x-2)(x-1)^2$

(Total 3 marks)

34 a) Expand and simplify $(x+1)(x-1)(x+3)$ **(3)**
  b) Hence expand $2x(x+1)(x-1)(x+3)$ **(1)**

(Total 4 marks)

35 a) Write $x^2-6x+13$ in the form $(x-a)^2+b$ **(2)**
  b) i) Find the minimum value of the expression $x^2-6x+13$
     ii) State the value of x for which this minimum value occurs. **(2)**

(Total 4 marks)

36 a) Write $2x^2+4x+7$ in the form $a(x+b)^2+c$ **(3)**
  b) Hence, or otherwise, write down the coordinates of the minimum point of the
     graph of $y=2x^2+4x+7$ **(2)**

(Total 5 marks)

37 a) Write $3-4x-x^2$ in the form $p-(x+q)^2$ **(2)**
  b) Write down the maximum value of the expression $3-4x-x^2$ **(1)**
  c) Write down the equation of the line of symmetry of the graph of
     $y=3-4x-x^2$ **(1)**

(Total 4 marks)

38 a) Given that $x^2 + 4x + 9 = (x + a)^2 + b$ for all values of x, find the value of a
and the value of b. **(2)**

b) Hence, or otherwise, explain why the graph with equation $y = x^2 + 4x + 9$
does not intersect the line with equation $y = 4$ **(1)**

(Total 3 marks)

39 a) Write $x^2 - 8x + 18$ in the form $(x - p)^2 + q$ **(2)**

b) Hence, or otherwise, explain why the equation $x^2 - 8x + 18 = 3$ has two solutions. **(1)**

(Total 3 marks)

40 a) Write $3x^2 - 6x + 1$ in the form $a(x + b)^2 + c$ **(3)**

b) Hence, or otherwise, find the range of values of k for which
$3x^2 - 6x + 1 = k$ has two positive solutions. **(2)**

(Total 5 marks)

41 Prove algebraically that the sum of the squares of any two consecutive integers is odd.

(Total 3 marks)

42 Prove algebraically that the difference between the squares of any two consecutive odd
numbers is a multiple of 8.

(Total 3 marks)

43 Prove algebraically that the sum of the squares of any two consecutive even numbers
leaves a remainder of 4 when divided by 8.

(Total 3 marks)

44 Prove algebraically that the sum of the squares of three consecutive positive integers is
always 1 less than a multiple of 3.

(Total 3 marks)

45 Prove algebraically that the sum of the squares of any three consecutive odd numbers is
never a multiple of 12.

(Total 3 marks)

46 Prove algebraically that, for any three consecutive integers, the difference between the
square of the largest integer and the square of the smallest integer is 4 times the middle
integer.

(Total 3 marks)

10 Expressions and formulae

1 Work out the value of $x^2 - 6x$ when $x = -3$

(Total 2 marks)

2 Work out the value of $6p - 2pq$ when $p = -3$ and $q = 4$

(Total 2 marks)

3 $u = \dfrac{fv}{v - f}$

Work out the value of u when $f = 14.08$ and $v = -6.4$

(Total 3 marks)

4 The formula for the area, A, of an ellipse is

$$A = \pi ab$$

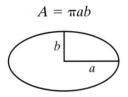

Calculate the value of a when $A = 24.6$ and $b = 1.7$
Give your answer correct to 1 decimal place.

(Total 2 marks)

5 $a = \dfrac{b}{c}$

Work out the value of a when $b = \dfrac{4}{9}$ and $c = \dfrac{5}{6}$

(Total 2 marks)

6 Simon repairs computers. The word formula gives the fee, in pounds, that Simon charges to repair a computer.

> Fee = £40 per hour that the repair takes plus £32 callout fee

Simon charges a fee of F pounds for a repair that takes H hours.
Write down a formula for F in terms of H.

(Total 2 marks)

7   Here is a pattern of shapes made from regular hexagons of side 1 cm:

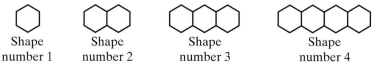

| Shape number 1 | Shape number 2 | Shape number 3 | Shape number 4 |

This rule can be used to find the perimeter, in cm, of a shape in this pattern.

> Double the Shape number
> then add 1
> and then multiply your answer by 2

P cm is the perimeter of Shape number n.
a)   Find the perimeter when the Shape number is 12   **(2)**
b)   Write down a formula for P in terms of n.   **(3)**
c)   Make n the subject of the formula in part **b)**.   **(3)**

(Total 8 marks)

8   Make r the subject of the formula $A = \pi r^2$

(Total 2 marks)

9   Make h the subject of the formula $y = m(x - h)$

(Total 3 marks)

10   Make s the subject of the formula $t = \sqrt{\dfrac{2s}{a}}$

(Total 3 marks)

11   Make x the subject of $3(x + 2b) = y(2 - 5x)$

(Total 4 marks)

12   Make x the subject of the formula $y = \dfrac{a}{x+a}$

(Total 4 marks)

13   Make r the subject of the formula $T = a(R + r)(R - r) + t$

(Total 4 marks)

14 A square with side of length L has area A and perimeter P.
A and P are given by the formulae

$$A = L^2$$
$$P = 4L$$

   a) Work out the value of P when $A = 121$ **(2)**
   b) Find a formula for P in terms of A.   **(2)**
   c) Find a formula for A in terms of P.   **(2)**

(Total 6 marks)

15 Make t the subject of $3s = \frac{1}{2}at^2 - 13$ where t is positive.

(Total 3 marks)

11 Linear equations and simultaneous equations

In all questions, you must show clear algebraic working.

1 Solve $4x - 1 = x + 7$

_____

(Total 2 marks)

2 Solve $8x + 3 = 2x - 2$

_____

(Total 3 marks)

3 Solve $2y + 7 = 6y + 5$

_____

(Total 3 marks)

4 Solve $5(x - 3) = 45$

_____

(Total 3 marks)

5 Solve $4(2x + 5) = 32$

_____

(Total 3 marks)

6 Solve $2(x + 4) = 9$

_____

(Total 3 marks)

7 Solve $5 - 8x = 8$

_____

(Total 3 marks)

8 Solve $9 - 4x = 7$

_____

(Total 3 marks)

9 Solve $\dfrac{5x + 3}{6} = 2$

_____

(Total 3 marks)

10 Solve $\dfrac{9 + 2x}{4} = 3x + 2$

_____

(Total 4 marks)

11 Solve $\dfrac{y+3}{4} + \dfrac{2y-3}{2} = 2$

(Total 4 marks)

12 Solve $4(3p + 1) = 15p + 8$

(Total 3 marks)

13 The diagram shows the lengths, in centimetres, of the sides of a triangle.

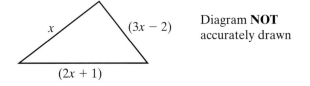

Diagram **NOT**
accurately drawn

The perimeter of the triangle is 14 cm.
a) Use this information to write an equation in x. **(2)**
b) Solve your equation. **(1)**
c) Find the length of the longest side of the triangle. **(1)**

(Total 4 marks)

14 Berto's age is x years. Rico's age is four times Berto's age.
In 10 years' time Rico's age will be twice Berto's age.
a) Use this information to write an equation in x. **(3)**
b) Solve your equation to find Berto's present age. **(3)**

(Total 6 marks)

15

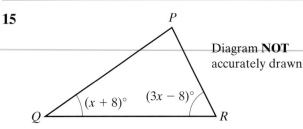

Diagram **NOT**
accurately drawn

PQR is a triangle.
Angle $Q = (x + 8)°$ and angle $R = (3x - 8)°$
The size of angle R is twice the size of angle Q.
i) Use this information to write an equation in x.
ii) Find the size of angle P.

(Total 6 marks)

16 $y = 2x - 5$

Work out the value of x when $y = -6$

(Total 2 marks)

17 $v = u + gt$

Find the value of t when $v = 16$, $u = 2$ and $g = 10$

(Total 2 marks)

18 $s = ut + \frac{1}{2}at^2$

Work out the value of a when $s = 60$, $u = 8$ and $t = 5$

(Total 2 marks)

19 Solve the simultaneous equations:

$y = x - 4$
$y = 9x$

(Total 3 marks)

20 Solve the simultaneous equations:

$4x + 3y = 3$
$2x - 6y = 9$

(Total 3 marks)

21 Solve the simultaneous equations:

$7x - 10y = 6$
$3x - 4y = 3$

(Total 4 marks)

22 a) Solve the simultaneous equations:

$$2x - 3y = 3$$
$$4x + 4y = 1$$

(3)

b) Write down the coordinates of the point of intersection of the two lines whose equations are:

$$2x - 3y = 3 \text{ and}$$
$$4x + 4y = 1$$

(1)

(Total 4 marks)

23 Work out the coordinates of the point of intersection of the line with equation $6x - 5y = 5$ and the line with equation $4x - 4y = 3$

(Total 4 marks)

12 Coordinates and graphs

1 Two points, A and B, are plotted on a grid.
A has coordinates $(0, 4)$ and B has coordinates $(6, 10)$.
Work out the coordinates of the midpoint of the line joining A and B.

(Total 2 marks)

2 P has coordinates $(-1, 2)$ and X has coordinates $(0, 5)$.
X is the midpoint of the line PQ.
Find the coordinates of point Q.

(Total 2 marks)

3

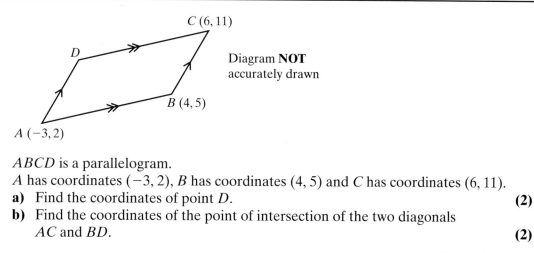

Diagram **NOT** accurately drawn

$ABCD$ is a parallelogram.
A has coordinates $(-3, 2)$, B has coordinates $(4, 5)$ and C has coordinates $(6, 11)$.
a) Find the coordinates of point D. **(2)**
b) Find the coordinates of the point of intersection of the two diagonals
AC and BD. **(2)**

(Total 4 marks)

4 Bishen sets off from home at 12 noon to meet his grandmother at the airport.
On the way, he discovers that he has left his mobile phone at home.
He returns home to collect it and sets off to the airport again to meet his grandmother.
At the airport he has to wait for her. He drives her the 10 km to her house and he then returns home.
The diagram on the next page shows Bishen's distance/time graph for the 80 minute journey.

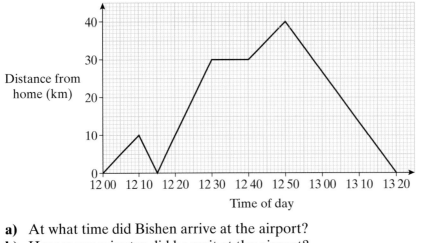

a) At what time did Bishen arrive at the airport? **(1)**
b) How many minutes did he wait at the airport? **(1)**
c) How many kilometres did Bishen travel between noon and 13 20? **(1)**
d) Between which two times was Bishen driving most quickly? **(1)**
e) At what times was Bishen 10 km from the airport? **(2)**
f) Calculate Bishen's speed during his journey home from his grandmother's house. Give your answer in kilometres per hour. **(2)**

(Total 8 marks)

5   Xanthe cycled from her home to visit her uncle.
Xanthe stopped at a shop on the way to buy him a present.
The diagram shows her distance/time graph for the journey to her uncle's house.

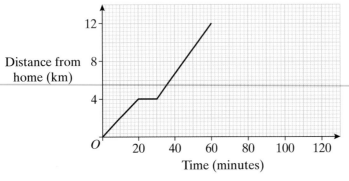

a) For how many minutes did Xanthe stop to buy the present? **(1)**
b) Find the distance from Xanthe's home to her uncle's house. **(1)**
c) Calculate the speed at which Xanthe cycled from home to the shop.
Give your answer in kilometres per hour. **(2)**
Xanthe stayed at her uncle's house for 30 minutes.
She then travelled back home at 48 km/h in her uncle's car.
d) i) Complete a copy of the distance/time graph.
ii) For how many minutes had Xanthe been away from home? **(4)**

(Total 8 marks)

6 The diagram shows the velocity/time graph for the first 40 seconds of a car journey.

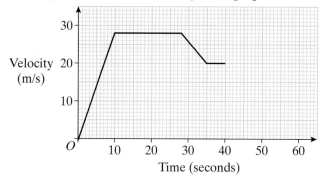

a) Find the constant acceleration of the car during the first 10 seconds. **(2)**

b) **i)** Write down the greatest velocity of the car in the first 40 seconds.

    ii) Calculate the distance travelled by the car while it is travelling at this greatest velocity. **(3)**

After 40 seconds the car decelerates at a constant rate of 2 m/s² and comes to rest.

c) Complete a copy of the velocity/time graph. **(2)**

(Total 7 marks)

___

7 A ball is projected vertically upwards from the edge of a flat roof.
t seconds after being projected, the height of the ball above the ground is h metres.
The diagram shows the graph of h against t, for $0 \leqslant t \leqslant 5$

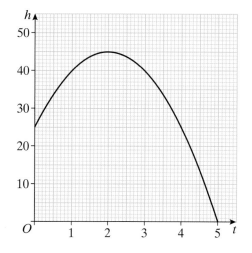

Using the graph, find:

a) the height above the ground from which the ball was projected **(1)**

b) the maximum height of the ball **(1)**

c) the total time for which the ball was at least 40 m above the ground. **(1)**

(Total 3 marks)

13 Linear graphs

1

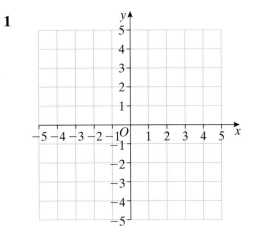

    a) On a copy of the grid, draw the line $x + y = 3$ **(1)**

    b) Find the coordinates of the point of intersection of the lines whose
equations are $x + y = 3$ and $x = -1$ **(2)**

                                                  (Total 3 marks)

2   **a)** On a copy of the grid, draw the line $y = \frac{1}{2}x - 2$ **(3)**

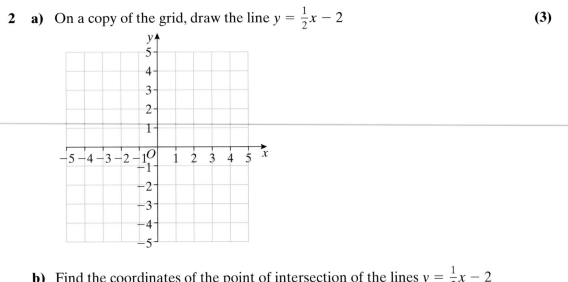

    b) Find the coordinates of the point of intersection of the lines $y = \frac{1}{2}x - 2$
and $y = -3$ **(2)**

                                                  (Total 5 marks)

3   On a copy of the grid, draw the graph of $y = 4x + 6$ from $x = -4$ to $x = 2$

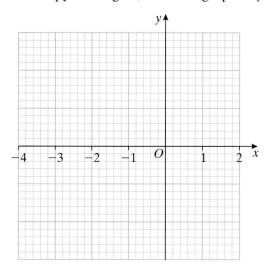

(Total 3 marks)

4   **a)**   On a copy of the grid, draw the line $x + y - 2 = 0$ **(1)**

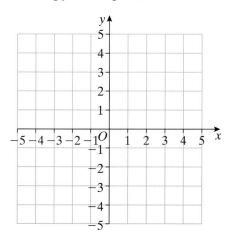

   b)   Write down an equation of a line parallel to the line $x + y - 2 = 0$ **(1)**

(Total 2 marks)

5 A ramp rises 80 cm in a horizontal distance of 280 cm.

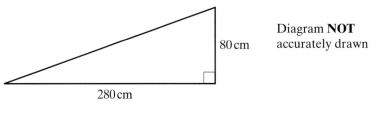

Diagram **NOT**
accurately drawn

80 cm

280 cm

Work out the gradient of the ramp.

(Total 2 marks)

6 The straight line, **L**, passes through the points $(0, -2)$ and $(3, 4)$.

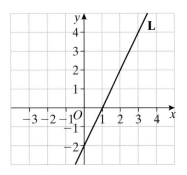

a) Work out the gradient of **L**. **(2)**

b) Write down an equation of **L**. **(2)**

(Total 4 marks)

7   The total monthly charge for Toni's phone is made up of a fixed fee for line rental plus the cost of the calls. The total monthly charge, £y, for a total of x minutes of calls can be found using the graph below for $x \leqslant 120$, using the straight line **L** on the grid.

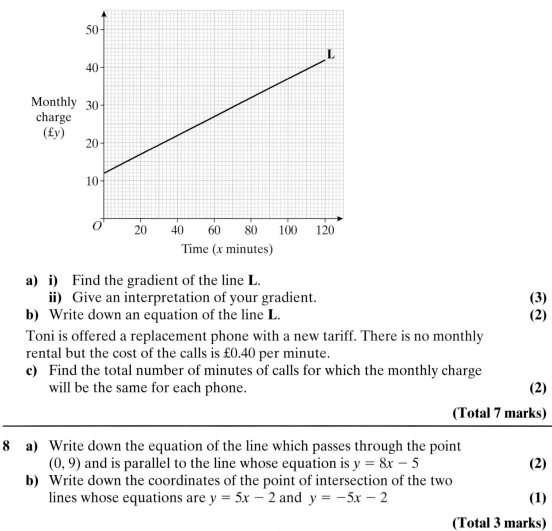

a) i)   Find the gradient of the line **L**.
    ii) Give an interpretation of your gradient.   **(3)**
b) Write down an equation of the line **L**.   **(2)**

Toni is offered a replacement phone with a new tariff. There is no monthly rental but the cost of the calls is £0.40 per minute.
c) Find the total number of minutes of calls for which the monthly charge will be the same for each phone.   **(2)**

(Total 7 marks)

8   a) Write down the equation of the line which passes through the point $(0, 9)$ and is parallel to the line whose equation is $y = 8x - 5$ **(2)**
    b) Write down the coordinates of the point of intersection of the two lines whose equations are $y = 5x - 2$ and $y = -5x - 2$ **(1)**

(Total 3 marks)

9

Find an equation of the straight line which passes through the points $(0, 7)$ and $(6, -2)$.

(Total 4 marks)

10 a) Find the gradient of the straight line **L** which passes through the points
$(1, 6)$ and $(5, -2)$. **(2)**

b) The line **M** is parallel to line **L** and intersects the y-axis at the point $(0, 1)$.
Write down an equation of **M**. **(2)**

c) The point $(3, k)$, where k is an integer, lies on the line **M**.
Find the value of k. **(1)**

(Total 5 marks)

11 Find the gradient of the line with equation $y - 3x = 8$

(Total 2 marks)

12 Find the gradient of the line with equation $4y + 3 = 2x$

(Total 2 marks)

13 Find the gradient of the line with equation $3x - 2y = 16$

(Total 3 marks)

14 The straight line, **L**, passes through the points $(4, 7)$ and $(-2, 3)$.

The straight line, **M**, passes through the points $(4, -3)$ and $(-2, 6)$.

Show that **L** and **M** are perpendicular to each other.

(Total 3 marks)

15 A is the point $(4, 5)$, B is the point $(-2, -3)$ and C is the point $(8, -1)$.

Find the equation of the line which is perpendicular to AB and passes through C.

Give your equation in the form $ax + by = c$ where a, b and c are integers.

(Total 4 marks)

16 Find the intercept on the y-axis of the straight line which passes through the point $(6, -3)$ and is perpendicular to the line with equation $y = -3x + 2$

(Total 3 marks)

17 A straight line, **L**, passes through the point $(3, 1)$ and is perpendicular to the line with equation $x + 2y = 6$

Find the equation of the straight line **L**.

(Total 3 marks)

18 Find the equation of the perpendicular bisector of the straight line joining the points $(-3, 4)$ and $(5, 2)$.

(Total 3 marks)

14 Sequences

1 The table shows the first three terms of a sequence:

Term number	1	2	3		
Term	8	15	24		

The rule for the sequence is

term $= (2 + \text{term number})^2 - 1$

a) Work out the next two terms of this sequence. **(2)**

b) 9999 is a term in this sequence.
Find the term number of this term. **(2)**

(Total 4 marks)

2 Here are the first five terms of a sequence:

  5   11   17   23   29

Find an expression, in terms of n, for the nth term of this sequence.

(Total 2 marks)

3 Here are the first four terms of an arithmetic sequence:

  42   38   34   30

Find an expression, in terms of n, for the nth term of this sequence.

(Total 2 marks)

4 The nth term of a sequence is given by this formula:

  nth term $= 62 - 5n$

a) Work out the 8th term of the sequence. **(1)**

b) Work out the sum of the 8th term and the 9th term of the sequence. **(2)**

c) Find an expression, in terms of n, for the sum of the nth term and $(n + 1)$th term of the sequence. Give your answer in its simplest form. **(2)**

d) Find the value of n for which $62 - 5n = -23$ **(2)**

(Total 7 marks)

5 Here are the first three terms of an arithmetic sequence:

$$-11 \quad -7 \quad -3$$

The kth term of this sequence is 169
Find the value of k.

(Total 4 marks)

6 a) Find an expression, in terms of n, for the nth term of the arithmetic sequence:

$$-2 \quad 6 \quad 14 \quad ...$$

(2)

b) The nth term of the arithmetic sequence $\quad -2 \quad 6 \quad 14 \quad ...$ is three times the nth term of the arithmetic sequence $\quad 151 \quad 152 \quad 153 \quad ...$
Find the value of n.

(3)

(Total 5 marks)

7 The first term of an arithmetic series is 6
The common difference of the series is 5
The series has 20 terms.
a) Find an expression, in terms of n, for the nth term of the series. **(1)**
b) Find the last term in the series. **(2)**
c) Find the sum of all the terms in the series. **(2)**

(Total 5 marks)

8 The first term of an arithmetic series is 18
The 8th term of the arithmetic series is 39
a) Find the common difference of the series. **(2)**
b) Find the sum of the first 40 terms of the series. **(2)**

(Total 4 marks)

9 Here is an arithmetic sequence with first term 2 and last term 149

$$2, \quad 5, \quad 8, \quad ..., \quad 149$$

a) Find the 10th term of this sequence. **(1)**
b) Find the number of terms in the sequence. **(2)**

(Total 3 marks)

10 Find the sum of all the odd numbers between 2 and 190

(Total 3 marks)

11 An arithmetic series has 45 terms.
The first term of the series is 60
The last term of the series is –6
Find the sum of all the terms in this arithmetic series.

(Total 3 marks)

12 The first term of an arithmetic series is 25
The sum of the first 16 terms of this arithmetic series is 160
Find the 16th term of this arithmetic series.

(Total 3 marks)

13 The second term of an arithmetic series is 120
The 12th term of this arithmetic series is 80
a) Find the common difference of the series. **(2)**
The sum of the first N terms of the arithmetic series is 0
b) Find the value of N. **(3)**

(Total 5 marks)

14 The 2nd term of an arithmetic series is 80
The 7th term of this arithmetic series is 20
a) Find the common difference of the series. **(2)**
b) Find the sum of all the positive terms of this arithmetic series. **(3)**

(Total 5 marks)

15 The first term of an arithmetic series is a.
The common difference of the series is d.
a) Write down an expression, in terms of a and d, for the sum of the first 3 terms
of the series. **(1)**
The sizes of the interior angles of triangle ABC form an arithmetic sequence.
b) Use your answer to part **a)** to show that one of the angles must be 60° **(2)**

(Total 3 marks)

16 The 12th term of an arithmetic series is 27
The 18th term of the arithmetic series is 45
a) Find the sum of the first 60 terms of this arithmetic series. **(5)**
b) Find the three consecutive terms in this arithmetic series whose sum is the
same as the sum of the first 60 terms of the series. **(3)**

(Total 8 marks)

17 The 9th term of an arithmetic series is $2\frac{1}{2}$

The sum of the 2nd term and the 5th term of this series is 27
Find the sum of the first 100 terms of the series.

(Total 6 marks)

18 An arithmetic series has first term a and common difference $2a$.
Show that the sum to $2N$ terms is always equal to four times the sum to N terms.

(Total 3 marks)

19 The sizes of the interior angles of a hexagon $ABCDEF$ form an arithmetic sequence.
The smallest interior angle of the hexagon $ABCDEF$ is $75°$
Find the size of the largest interior angle of the hexagon $ABCDEF$.

(Total 3 marks)

20 a) Find the sum of the arithmetic series whose first term is 12, common
difference is 3 and whose last term is 99 **(3)**
   b) Find the sum of all 2-digit numbers. **(2)**
   c) Hence find the sum of all 2-digit numbers which are **not** multiples of 3 **(1)**

(Total 6 marks)

15 Quadratic equations

In all questions, you must show clear algebraic working.

1 The width of a rectangle is $(x + 2)$ cm. The perimeter of the rectangle is $6(x + 1)$ cm. The area of the rectangle is 18 cm^2.
Show that $2x^2 + 5x - 16 = 0$

(Total 3 marks)

2 Solve $x^2 - 4x = 0$

(Total 2 marks)

3 Solve $y^2 - 3y - 108 = 0$

(Total 3 marks)

4 Solve $2x^2 - 5x - 3 = 0$

(Total 3 marks)

5 Solve $\dfrac{3(3x - 4)}{x + 2} = x - 2$

(Total 4 marks)

6 Solve the equation $x^2 + 7x + 8 = 0$
Give your answers correct to 3 significant figures.

(Total 3 marks)

7 Solve the equation $5y^2 - 2y - 4 = 0$
Give your answers correct to 3 significant figures.

(Total 3 marks)

8 $y = p^2 + 6p$
Find the values of p when $y = 3$
Give your answers correct to 2 decimal places.

(Total 3 marks)

9 Solve the simultaneous equations
$$y = 2x^2$$
$$y = 3x + 9$$

(Total 6 marks)

10 Solve the simultaneous equations:

$$y = 2x + 3$$
$$x^2 + y^2 = 5$$

(Total 6 marks)

11 Solve the simultaneous equations:

$$2x + y = 1$$
$$x^2 + y^2 = 13$$

(Total 7 marks)

12 Solve the simultaneous equations:

$$x = 2y^2 - 4$$
$$y = \tfrac{1}{2}x$$

(Total 6 marks)

13 A square has sides of length $(x + 2)$ cm.
A right-angled isosceles triangle has its two equal sides of length $(2x + 1)$ cm.
The area of the square is equal to the area of the triangle.

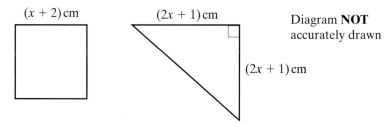

$(x + 2)$ cm $\qquad$ $(2x + 1)$ cm $\qquad$ Diagram **NOT** accurately drawn

$(2x + 1)$ cm

    a) Write down an equation in x. **(1)**
    b) Show that your equation simplifies to $2x^2 - 4x - 7 = 0$ **(2)**
    c) By solving the equation $2x^2 - 4x - 7 = 0$, find the perimeter of the square.
        Give your answer correct to 3 significant figures. **(4)**

(Total 7 marks)

14 a) Express $x^2 - 4x + 1$ in the form $(x + a)^2 + b$ **(2)**

    b) Hence solve the equation $x^2 - 4x + 1 = 0$ giving your answer in the
        form $p \pm \sqrt{q}$ where p and q are integers. **(2)**

(Total 4 marks)

16 Inequalities

1   Solve the inequality $2x + 7 > 13$

_____

(Total 2 marks)

2   Solve the inequality $7x - 3 < 3x + 7$

_____

(Total 3 marks)

3   Solve the inequality $-1 \leqslant x + 1 < 4$

_____

(Total 2 marks)

4   Solve the inequality $-2 < 2x \leqslant 6$

_____

(Total 2 marks)

5   Solve the inequality $0 < 2x + 3 < 7$

_____

(Total 3 marks)

6   **a)** Solve the inequality $2x + 7 > 1$ **(2)**
   b) On a copy of the number line, represent the solution to part **a)**.

(2)

(Total 4 marks)

7   **a)** Solve the inequality $2x + 5 \leqslant 11$ **(2)**
   b) On a copy of the number line, represent the solution to part **a)**.

(2)

(Total 4 marks)

8   **a)** Solve the inequality $-2 \leqslant 4x + 6 < 22$ **(3)**
   b) On a copy of the number line, represent the solution to part **a)**.

(2)

(Total 5 marks)

9 **a)** Solve the inequality $2x + 3 < 13$ **(2)**

b) n is a positive integer.
Write down all the values of n which satisfy the inequality $2n + 3 < 13$ **(2)**

(Total 4 marks)

10 **a)** Solve the inequality $4x + 17 > 3$ **(2)**

b) n is a negative integer.
Write down all the values of n which satisfy the inequality $4n + 17 > 3$ **(2)**

(Total 4 marks)

11 **a)** Solve the inequality $-4 \leqslant 2x < 3$ **(2)**

b) n is an integer.
Write down all the values of n which satisfy the inequality $-4 \leqslant 2n < 3$ **(2)**

(Total 4 marks)

12 Show, by shading on a copy of the grid, the region which satisfies both these inequalities:

$$2 \leqslant x \leqslant 4 \quad \textbf{and} \quad -3 \leqslant y \leqslant -1$$

Label your region **R**.

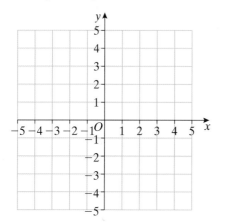

(Total 3 marks)

13 On a copy of the grid, show clearly the region defined by the inequalities:

$$x + y \leqslant 3 \qquad x \geqslant -1 \qquad y > 2$$

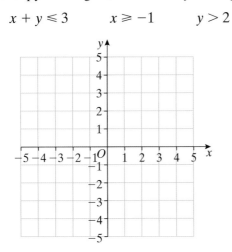

(Total 4 marks)

14 Show, by shading on a copy of the grid, the region which satisfies all three of these inequalities:

$$y \leqslant 4 \qquad y \leqslant 4x - 2 \qquad y \geqslant 2x$$

Label your region **R**.

(Total 4 marks)

15 Show, by shading on a copy of the grid, the region which satisfies all three of these inequalities:

$$1 \leqslant y \leqslant 4 \qquad y \geqslant -2x \qquad y + 2x \leqslant 4$$

Label your region **R**.

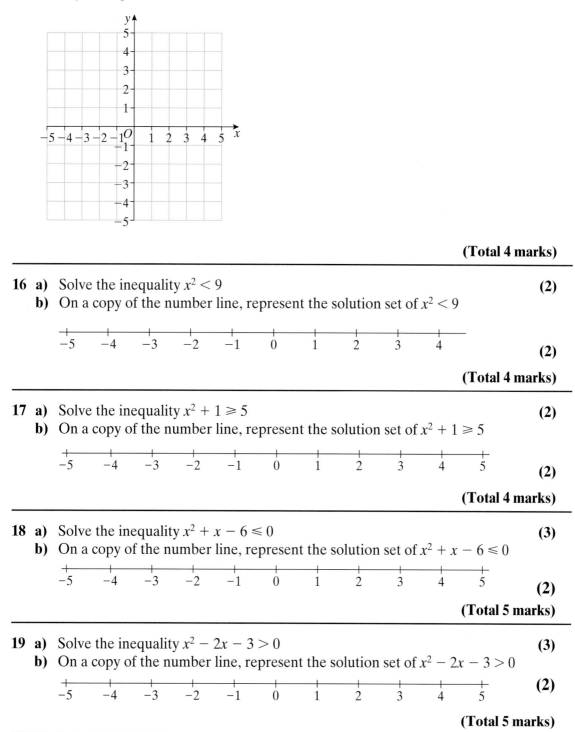

(Total 4 marks)

---

16 **a)** Solve the inequality $x^2 < 9$ **(2)**
    b) On a copy of the number line, represent the solution set of $x^2 < 9$

(2)

(Total 4 marks)

---

17 **a)** Solve the inequality $x^2 + 1 \geqslant 5$ **(2)**
    b) On a copy of the number line, represent the solution set of $x^2 + 1 \geqslant 5$

(2)

(Total 4 marks)

---

18 **a)** Solve the inequality $x^2 + x - 6 \leqslant 0$ **(3)**
    b) On a copy of the number line, represent the solution set of $x^2 + x - 6 \leqslant 0$

(2)

(Total 5 marks)

---

19 **a)** Solve the inequality $x^2 - 2x - 3 > 0$ **(3)**
    b) On a copy of the number line, represent the solution set of $x^2 - 2x - 3 > 0$

(2)

(Total 5 marks)

---

20 Solve the inequality $2x - x^2 \geqslant 0$

(Total 3 marks)

21 a) Solve the inequality $2x^2 - 3x - 2 > 0$ **(3)**
  b) Write down the smallest positive integer which satisfies this inequality. **(1)**

(Total 4 marks)

22 a) Solve the inequality $6x^2 - 19x + 10 < 0$ **(3)**
  b) n is an integer.
    Write down all the values of n which satisfy the inequality
    $6n^2 - 19n + 10 < 0$ **(1)**

(Total 4 marks)

17 Indices

1 **a)** Simplify $a \times a \times a \times a \times a \times a \times a$ **(1)**
   b) Simplify $p^2 \times p^4 + 3p \times p^5$ **(2)**

   c) Simplify $\dfrac{q^6}{q^2}$ **(1)**

                                                                        (Total 4 marks)

2 **a)** Work out the value of $2y^3$ when $y = 2$ **(1)**
   b) Simplify:
      i) $c^4 \times c^5$
      ii) $d^6 \div d$ **(2)**

                                                                        (Total 3 marks)

3 **a)** Work out the value of $x^2 + 5x$ when $x = -3$ **(2)**

   b) Simplify $\dfrac{t^5 \times t^6}{t^7}$ **(1)**

                                                                        (Total 3 marks)

4 **a)** $F = ab^2 - b^3$
      Work out the value of F when $a = 3$ and $b = -2$ **(2)**

   b) Simplify $p^6 q^2 \times 2pq^3$ **(2)**

   c) Simplify $\dfrac{3r \times (2r)^4}{r^3}$ **(2)**

   d) Evaluate $5x^0$ when $x = 2$ **(1)**

                                                                        (Total 7 marks)

5 **a)** Complete a copy of the table of values
      for $y = x^2 + 2x - 2$

x	-4	-3	-2	-1	0	1	2
y			-2		-2		6

                                                                                           (2)

   b) On a copy of the grid, draw the
      graph of $y = x^2 + 2x - 2$

                                                                                           (2)

                                                                        (Total 4 marks)

6 Simplify:

   a) $(\sqrt{x})^6$

   b) $\dfrac{16y^4}{(2\sqrt{y})^4}$

(Total 3 marks)

7 **a)** Simplify $(3a^4)^3$ **(2)**

   b) Simplify $(8p^9)^{\frac{1}{3}}$ **(2)**

(Total 4 marks)

8 **a)** Simplify $x^6 \times (x^2)^{-3}$ **(2)**

   b) Simplify $(2p^3q^4)^4$ **(2)**

   c) Simplify $(r^3t^{-4})^{-2}$ **(2)**

(Total 6 marks)

9 Simplify:

   a) $(64x^{12})^{\frac{2}{3}}$ **(2)**

   b) $\left(\dfrac{x^2}{4}\right)^{-\frac{1}{2}}$ **(2)**

(Total 4 marks)

18 Proportion

1 The extension, E cm, of a spring is directly proportional to the tension, T newtons.
When $T = 14$, $E = 3.5$
   a) Express E in terms of T. **(3)**
   b) Find the value of E when $T = 10$. **(1)**
   c) Calculate the tension when the extension is 6.5 cm. **(1)**

   (Total 5 marks)

2 The area, A cm², of a square is directly proportional to the square of the length, D cm, of its diagonal.
When $D = 7$, $A = 24.5$
   a) i) Express A in terms of D.
   ii) On a copy of the axes, sketch the graph of A against D:

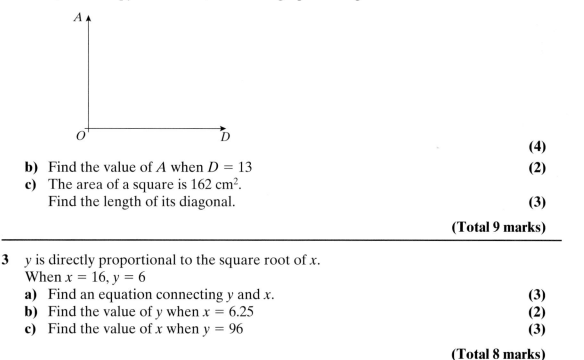

   (4)
   b) Find the value of A when $D = 13$. **(2)**
   c) The area of a square is 162 cm².
   Find the length of its diagonal. **(3)**

   (Total 9 marks)

3 y is directly proportional to the square root of x.
When $x = 16$, $y = 6$
   a) Find an equation connecting y and x. **(3)**
   b) Find the value of y when $x = 6.25$ **(2)**
   c) Find the value of x when $y = 96$ **(3)**

   (Total 8 marks)

4 The mass, M kg, of a solid metal hemisphere is directly proportional to the cube of the radius, r cm, of its circular base.
When $r = 10$, $M = 16$
   a) Express M in terms of r. **(3)**
   b) The radius of the base of a hemisphere is 7.5 cm.
   Calculate its mass. **(2)**
   c) The mass of a hemisphere is 250 kg.
   Calculate the **diameter** of its circular base. **(3)**

(Total 8 marks)

5 For a fixed mass of gas, the volume, V m^3, varies inversely as the pressure, P newtons/m^2.
When $V = 5$, $P = 200$
   a) i) Express V in terms of P.
   ii) On a copy of the axes, sketch the graph of V against P:

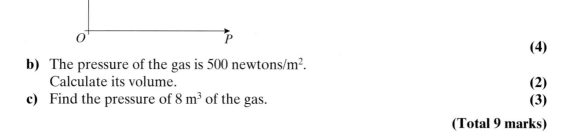

(4)

   b) The pressure of the gas is 500 newtons/m^2.
   Calculate its volume. **(2)**
   c) Find the pressure of 8 m^3 of the gas. **(3)**

(Total 9 marks)

6

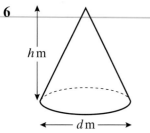

The diameter, d m, of a set of cones, each with the same volume, is inversely proportional to the square root of the height, h m.
When $h = 2.56$, $d = 1.5$
   a) Express d in terms of h. **(3)**
   b) Find d when $h = 10.24$ **(2)**
   c) Find h when $d = 3$ **(3)**

(Total 8 marks)

7   The force, F, between two magnets is inversely proportional to the square of the distance, D, between them.
When $D = 4$, $F = 2$
   a) Find a formula for F in terms of D.   (3)
   b) Find the value of F when $D = 5$ (2)
   c) Find the value of D when $F = 12.5$ (3)

                                                                    (Total 8 marks)

---

8   A pebble is dropped from the edge of a cliff.
The edge of the cliff is 250 metres above the sea.
After t seconds, the pebble has fallen a distance x metres.
x is directly proportional to t^2
When $t = 2$, $x = 19.6$
   a) Find an equation connecting x and t.   (3)
   b) Find the value of x when $t = 5$ (2)
   c) Find how long the pebble takes to fall the final 73.6 m before landing in the sea.   (4)

                                                                    (Total 9 marks)

---

9 T is inversely proportional to the square root of h.
When $h = 25$, $T = 8$
   a) Find a formula for T in terms of h.   (3)
   b) Work out the value of T when $h = 64$ (2)
   c) Work out the value of h when $T = 80$ (2)

                                                                    (Total 7 marks)

---

10 The time, T seconds, for a pendulum to swing is inversely proportional to the square root of the acceleration due to gravity, g m/s^2
On Earth, $g = 9.8$ and a pendulum takes 1 second to swing.
   a) Find a formula for T in terms of g.   (2)
On the Moon, $g = 1.6$
   b) Calculate the time the pendulum takes to swing on the Moon.
           Give your answer correct to 3 significant figures.   (2)
On Mars, the pendulum takes 1.63 s to swing.
   c) Calculate the acceleration due to gravity on Mars.
           Give your answer correct to 3 significant figures.   (2)

                                                                    (Total 6 marks)

---

11 C is inversely proportional to v^3
$C = 40$ when $v = 1.5$
   a) Express C in terms of v.   (3)
   b) Calculate the value of v when $C = 2.56$ (2)

                                                                    (Total 5 marks)

---

12 The number, N, of identical spheres which can be made from a given weight of metal varies inversely as the cube of the radius, r millimetres, of each sphere.
500 spheres of radius 2 mm can be made from a certain weight of metal.

a) Find a formula for N in terms of r. **(3)**

b) Calculate the number of spheres of radius 5 mm which can be made from the same weight of metal. **(2)**

(Total 5 marks)

19 Function notation and transformation of functions

1 The function f is defined as $f(x) = \dfrac{x+1}{2}$

    a) Find f(7)   **(1)**

    b) Solve $f^{-1}(x) = 2$ **(2)**

                                               (Total 3 marks)

2 $f(x) = 4 - 3x$

       $g(x) = x + 1$

    a) Find $fg(x)$ **(2)**

    b) Find $gf(x)$ **(2)**

                                               (Total 4 marks)

3 $f : x \mapsto \dfrac{3}{x+2}, \; x \ne -2$

    Find the inverse function f^{-1} in the form: $f^{-1} : x \mapsto \ldots$

                                               (Total 3 marks)

4 The functions f and g are defined as follows:

    $f(x) = \dfrac{2}{x+4}$

    $g(x) = \sqrt{x+5}$

    Calculate $fg(11)$

                                               (Total 3 marks)

5 The diagram shows part of the graph of $y = f(x)$.

Use the graph to:

    a) find f(2)   **(1)**

    b) solve $f(x) = 2$ **(2)**

    c) find ff(0.5)   **(2)**

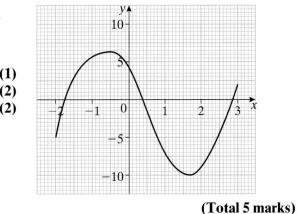

                                               (Total 5 marks)

6   f and g are functions:

$$f : x \mapsto 4x - 1$$
$$g : x \mapsto \sqrt{(x+1)}$$

a) Calculate f(−2)   **(2)**
b) Given that f(a) = 5, find the value of a.   **(2)**
c) Calculate gf(9)   **(2)**
d) Which values of x cannot be included in the domain of g?   **(1)**
e) Find the inverse function g^{-1} in the form g$^{-1} : x \mapsto \ldots$ **(3)**

(Total 10 marks)

7   The function f is defined as $f(x) = \dfrac{x+3}{x-1}$

a) Find the value of:
  i)   f(5)

  ii) $f\left(\tfrac{1}{2}\right)$ **(2)**

b) State which value(s) of x must be excluded from the domain of f.   **(1)**
c) **i)**   Find ff(x). Give your answer in its simplest form.
    ii) What does your answer to **c i)** show about the function f?   **(4)**

(Total 7 marks)

8       $f(x) = 3 - 2x$
        $g(x) = x^2$

Solve the equation $f^{-1}(x) = gf(x)$

(Total 6 marks)

9       $f : x \mapsto 4x - 3$

        $g : x \mapsto \dfrac{4}{x}, \; x \neq 0$

a) Find the value of:
  i)   f(0)
  ii) fg(16)   **(2)**
b) Express the inverse function f^{-1} in the form f$^{-1} : x \mapsto \ldots$ **(2)**
c) **i)**   Express the composite function gf in the form gf $: x \mapsto \ldots$
    ii) Which value of x must be excluded from the domain of gf?   **(2)**

(Total 6 marks)

10 f and g are functions.

$$f(x) = \frac{3x+2}{4} \qquad fg(x) = x \qquad gf(x) = x$$

   a) Find $g(x)$ **(3)**

   b) Solve the equation $g^{-1}(x) = x + 2$ **(3)**

                                                          (Total 6 marks)

---

11 With equal scales on the x-axis and y-axis, the graph of $y = f^{-1}(x)$ can be obtained by reflecting the graph of $y = f(x)$ in the line **L**.

   a) Write down the equation of the line **L**.   **(1)**

   b) Given that $f(x) = \dfrac{4x+5}{3}$, solve the equation $f(x) = f^{-1}(x)$ **(3)**

                                                            (Total 4 marks)

---

12 $f : x \mapsto x^2 + 2x + 3, \ x \geqslant 0$

   a) Write $x^2 + 2x + 3$ in the form $(x + c)^2 + d$ **(2)**

   b) The domain of f is $x \geqslant 0$ Find the range of f.   **(1)**

   c) Find the inverse function f^{-1} in the form $f^{-1} : x \mapsto \dots$ **(3)**

                                                            (Total 6 marks)

---

13 The function f is defined for all values of x as $f : x \mapsto 2x^2 - 10x - 5$

   a) Write $2x^2 - 10x - 5$ in the form $a(x + b)^2 + c$ **(3)**

   b) Find the range of f.   **(1)**

                                                             (Total 4 marks)

---

14 $g : x \mapsto 3x^2 + 6x + 10, \ x \geqslant 0$

   Find the inverse function g^{-1} in the form $g^{-1} : x \mapsto \dots$

                                                            (Total 6 marks)

---

15 A single transformation maps the curve with equation $y = \sin x°$ onto the curve with equation $y = 5 + \sin x°$

   Describe the transformation.

                                                            (Total 2 marks)

---

16 The graph of $y = x^2 - 3x + 1$ is reflected in the x-axis.

   Find the equation of the new graph.

                                                            (Total 2 marks)

---

17 The graph of $y = x^2 - 5x + 3$ is reflected in the y-axis.

   Find the equation of the new graph.

                                                            (Total 2 marks)

---

18 The graph of $y = \sin x°$ is translated by the vector $\begin{pmatrix} 30 \\ 0 \end{pmatrix}$.

   Write down the equation of the new graph.

                                                            (Total 2 marks)

19 Describe fully the single transformation which maps the graph of $y = \cos x°$ onto the graph of $y = 2 + \cos(x + 60)°$

(Total 2 marks)

20 A single transformation maps the curve with equation $y = x^2 - 2x + 4$ onto the curve with equation $y = x^2 + 2x + 4$
Describe the transformation.

(Total 2 marks)

21

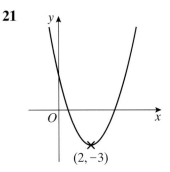

(2, −3)

The diagram shows part of the curve with equation $y = f(x)$
The curve has one turning point whose coordinates are $(2, -3)$
a) Write down the coordinates of the turning point of the curve with equation
   i) $y = f(x) + 6$
   ii) $y = -f(x)$
   iii) $y = 5f(x)$
   iv) $y = f(x - 4)$
   v) $y = f(2x)$

(5)

A translation is applied to the curve with equation $y = f(x)$ to give the curve with equation $y = g(x)$. The curve $y = g(x)$ has a turning point at the origin O.
b) Write down the translation vector.

(1)

(Total 6 marks)

20 Harder graphs

1 **a)** Complete a copy of the table of values for $y = x^3 + 3x^2 - 3$

x	-4	-3	-2	-1	0	1	2
y		-3					

(2)

b) On a copy of the grid, draw the graph of $y = x^3 + 3x^2 - 3$

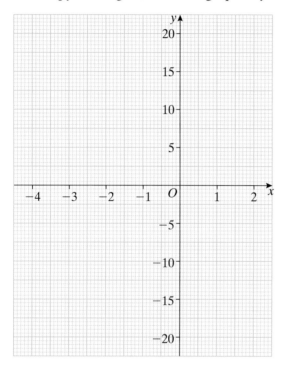

(2)

c) Use your graph to find estimates, correct to 1 decimal place where appropriate, for the solutions of:

i) $x^3 + 3x^2 - 3 = 0$

ii) $x^3 + 3x^2 - 4 = 0$

(4)

(Total 8 marks)

2 Part of the graph of $y = 6 - 2x - x^2$ is shown on the grid.

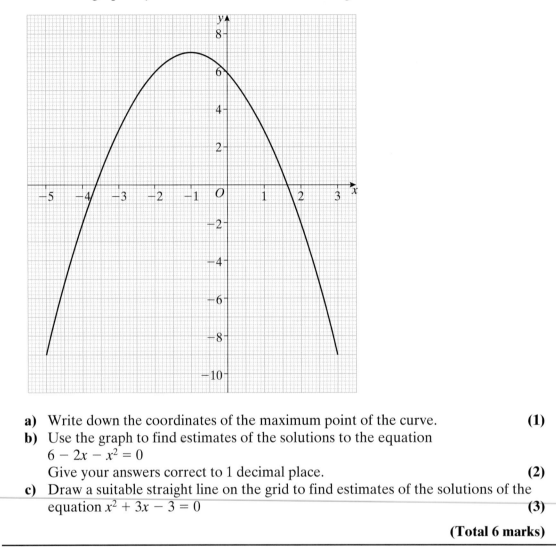

a) Write down the coordinates of the maximum point of the curve. **(1)**

b) Use the graph to find estimates of the solutions to the equation
$6 - 2x - x^2 = 0$
Give your answers correct to 1 decimal place. **(2)**

c) Draw a suitable straight line on the grid to find estimates of the solutions of the
equation $x^2 + 3x - 3 = 0$ **(3)**

(Total 6 marks)

3 **a)** Complete a copy of the table of values for $y = x + \dfrac{4}{x}$

x	0.4	0.6	0.8	1	1.5	2	3	4	5	6
y		7.3		5	4.2		4.3		5.8	6.7

(2)

b) On a copy of the grid, draw the graph of $y = x + \dfrac{4}{x}$ for $0.4 \leqslant x \leqslant 6$

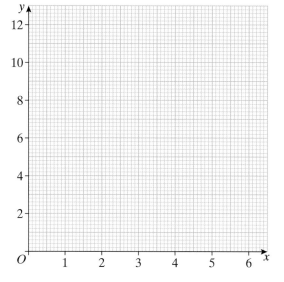

(2)

c) Use your graph to find estimates for the solutions of the equation

$$x + \frac{4}{x} = 6$$

(2)

The solutions of the equation $3x + \dfrac{4}{x} = 10$ are the x-coordinates of the points of

intersection of the graph $y = x + \dfrac{4}{x}$ and a straight line **L**.

d) Find an equation of **L**.

(2)

(Total 8 marks)

4 Part of the graph of $y = x^3 - 4x$ is shown on the grid.

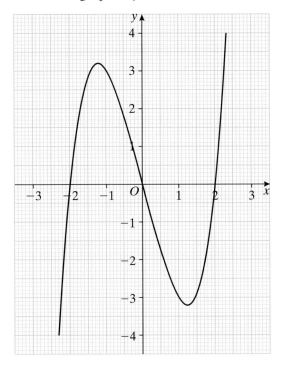

The graph of $y = x^3 - 4x$ and the line with equation $y = k$, where k is an integer, have three points of intersection.

a) Find the least possible value of the integer k. **(1)**

b) By drawing a suitable straight line on the grid, find estimates of the solutions of the equation $x^3 - 3x + 1 = 0$

Give your answers correct to 1 decimal place. **(4)**

(Total 5 marks)

5 The diagram shows part of the graph of $y = f(x)$:

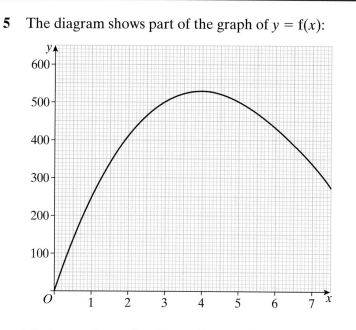

Find an estimate for the gradient of the curve at the point where $x = 2$

(Total 3 marks)

6 On a copy of the grid, draw the graph of $y = 10 + 10x + \dfrac{480}{x^2}$ for $1 \leqslant x \leqslant 7$

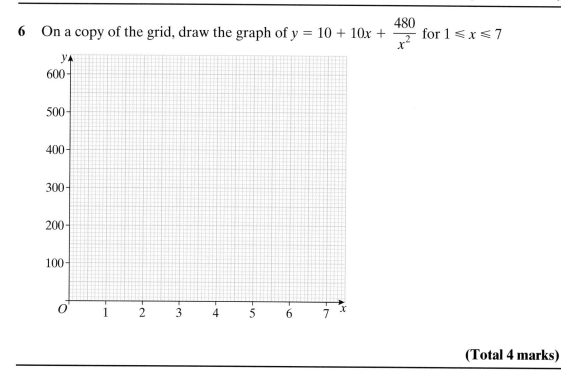

(Total 4 marks)

7 Here are three graphs.

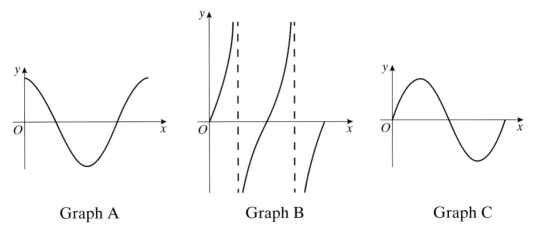

Graph A   Graph B   Graph C

Copy and complete the table below with the letter of the graph that could represent each given equation.

Equation	Graph
$y = \sin x$	
$y = \cos x$	
$y = \tan x$	

(Total 2 marks)

8 **a)** Complete a copy of the table of values for $y = 3 \sin x°$

Where necessary, give values correct to 2 significant figures.

x	0	30	60	90	120	150	180	210	240	270	300	330	360
y			2.6								−2.6		

(2)

b) On a copy of the grid, draw the graph of $y = 3 \sin x°$ for $0 \le x \le 360$

(2)

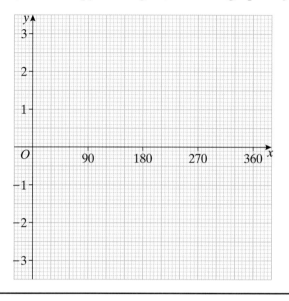

(Total 4 marks)

9 Here is a sketch of the graph of $y = \cos 2x°$ for $0 \le x \le 45$

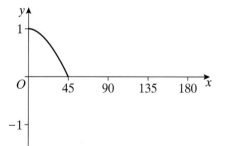

On a copy of the diagram, complete a sketch of the graph of $y = \cos 2x°$
for $0 \le x \le 180$

(Total 3 marks)

21 Calculus

1 For the curve $y = 4x^2 - 4x + 1$:

a) find $\dfrac{dy}{dx}$

b) find the gradient of the curve at the point where $y = 0$

(Total 6 marks)

---

2 A curve has equation $y = 4x^2 + \dfrac{1}{x}$

The curve has one stationary point.

Find $\dfrac{dy}{dx}$ and use your answer to find the coordinates of this stationary point.

(Total 4 marks)

---

3 A particle moves along a line.
For $t \geqslant 1$, the distance of the particle from O at time t seconds is x metres, where

$$x = \frac{9}{2t} + t$$

Find the acceleration of the particle when $t = 3$

(Total 4 marks)

---

4 The curve $y = \dfrac{x^2}{3} - 4x + 7$ crosses the y-axis at the point P.

Q is the point $(6, -5)$.
a) Find the gradient of the line PQ.   **(3)**

b) For the curve, find $\dfrac{dy}{dx}$ **(2)**

The gradient of the curve at point R is equal to the gradient of the line PQ.
c) Show that R is vertically below the midpoint of the line PQ.   **(3)**

(Total 8 marks)

5   A body is moving in a straight line which passes through a fixed point O.
    The displacement, x metres, of the body from O at time t seconds is given by

$$x = t^3 + 3t^2 - 9t$$

   a) Find an expression for the velocity, v m/s, at time t seconds.   **(2)**
   b) Find the acceleration when the body is instantaneously at rest.   **(4)**

   (Total 6 marks)

6   A curve has equation $y = 7 + 6x - 2x^2$
   a) For this curve find:

   i) $\dfrac{dy}{dx}$

   ii) the coordinates of the turning point.   **(4)**

   b) State, with a reason, whether the turning point is a maximum or a minimum.   **(2)**
   c) Find the equation of the line of symmetry of the curve $y = 7 + 6x - 2x^2$ **(2)**

   (Total 8 marks)

7   The diagram shows the graph of $y = x^3 - 3x^2 - 9x + 8$
    A is the maximum point on the curve.
    B is the minimum point on the curve.

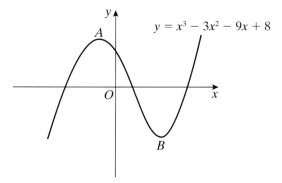

   a) For the equation $y = x^3 - 3x^2 - 9x + 8$

   i)   find $\dfrac{dy}{dx}$

   ii) find the x coordinates of A and B.   **(6)**

   The line $y = k$, where k is a number, is a tangent to the curve $y = x^3 - 3x^2 - 9x + 8$
   b) Find the two possible values of k.   **(2)**

   (Total 8 marks)

8 The owner of 40 apartments knows that, at a rent of $360 per week, all apartments will be occupied. For each increase in rent of $30 per week, one fewer apartment will be occupied.

He increases the rent from $360 per week to $(360 + 30x)$ per week.

After this increase, the total income he receives in rent is $R per week.

a) Show that $R = 14\,400 + 840x - 30x^2$ **(3)**

b) Find $\dfrac{\mathrm{d}R}{\mathrm{d}x}$ **(2)**

c) What weekly rent should he charge in order to achieve the maximum income? **(3)**

(Total 8 marks)

---

9 A curve has equation $y = x^3 + 75x - 100$

a) For the curve, find $\dfrac{\mathrm{d}y}{\mathrm{d}x}$ **(2)**

b) Find the coordinates of the points on the curve where the gradient is 87 **(3)**

c) Use your answer to part **a)** to determine the number of turning points the curve has. **(2)**

(Total 7 marks)

---

10 The length of a cuboid is twice its width.

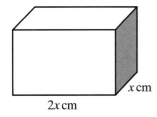

x cm

$2x$ cm

The volume of the cuboid is 72 cm³.

The surface area of the cuboid is A cm².

a) Show that $A = 4x^2 + \dfrac{216}{x}$ **(3)**

b) Find $\dfrac{\mathrm{d}A}{\mathrm{d}x}$ **(2)**

c) Find the height of the cuboid for which A is a minimum. **(3)**

(Total 8 marks)

11

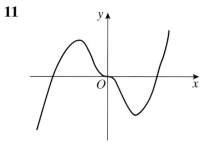

The diagram shows a sketch of the curve with equation $y = 3x^5 - 20x^3$

a) Find the x-coordinates of the stationary points. **(4)**

b) Find the coordinates of the turning points. **(2)**

(Total 6 marks)

Shape, Space and Measures

22 Compound measures

1 Rajesh cycled 60 km in 2 hours 30 minutes.
Work out his average speed in km/h.

(Total 3 marks)

2 Sita drove 57 km in 45 minutes.
Work out her average speed in km/h.

(Total 3 marks)

3 Ben ran 100 metres at an average speed of 8.6 m/s.
Work out his time.
Give your answer correct to 2 decimal places.

(Total 3 marks)

4 Preety drove 108 km at an average speed of 45 km/h.
Work out the time it took her.
Give your answer in hours and minutes.

(Total 3 marks)

5 Ali walked for 3 hours 40 minutes at an average speed of 5.4 km/h.
Work out the distance he walked.

(Total 3 marks)

6 Jasmin flew from Dubai to Hong Kong.
Her flight covered a distance of 6052 km.
The flight time was 8 hours 30 minutes.
Work out the average speed for the flight.

(Total 3 marks)

7 Lee flew from Nairobi to Cairo.
The flight time was 4 hours 50 minutes.
The average speed was 738 km/h.
Work out the distance from Nairobi to Cairo.

(Total 3 marks)

8   Tara flew from Amsterdam to Kuala Lumpur.
     Her flight covered a distance of 10 430 km.
     The average speed was 840 km/h.
     Work out the flight time.
     Give your answer in hours and minutes.

     (Total 3 marks)

9   A snail moves 90 cm in $2\frac{1}{2}$ minutes.
     Work out its average speed.
     Give your answer in m/s.

     (Total 3 marks)

10 The distance between Delhi and Mumbai is 1411 km.
     A train covers this distance at an average speed of 85 km/h.
     Work out the time it takes.
     Give your answer in hours and minutes.

     (Total 3 marks)

11 An iron cylinder has a radius of 12 cm and a length of 18 cm.
     The mass of the cylinder is 64 kg.
     Work out the density of iron
     i)   in g/cm^3
     ii) in kg/m^3
     Give your answers correct to 3 significant figures.

     (Total 4 marks)

12 The volume of a gold bar is 51.8 cm^3
     The density of gold is 19.3 g/cm^3
     Calculate the mass of a gold bar.
     Give your answer to the nearest gram.

     (Total 2 marks)

13 The mass of a pine plank is 5.56 kg.
     The density of pine is 373 kg/m^3
     Find the volume of the plank.
     Give your answer in m^3, correct to 3 significant figures.

     (Total 2 marks)

14 $\text{Pressure} = \dfrac{\text{Force}}{\text{Area}}$

where Pressure is in Pascals, Force is in Newtons and Area is in m²
The lengths of the edges of a rectangular block are 1.5 m, 2 m and 2.5 m.
The weight of the block is 7500 N.
The box rests with one of its faces in contact with the ground.
Calculate the least possible pressure the box exerts on the ground.

(Total 3 marks)

15 $P = \dfrac{F}{A}$

where P Pascals is Pressure, F Newtons is Force and A m² is Area.
A force exerts a pressure of 250 Pa on a circular area.
The radius of the circular area is 1.8 m.
Find the size, in Newtons, of the force.
Give your answer correct to 3 significant figures.

(Total 3 marks)

16 $\text{Pressure} = \dfrac{\text{Force}}{\text{Area}}$

where Pressure is in Newtons/m², Force is in Newtons and Area is in m²
An elephant has a weight of 48 000 Newtons.
When standing on all four feet, it exerts a pressure of 600 000 Newtons/m² on the ground.
Work out the area, in m², of each of its feet.

(Total 3 marks)

17 $\text{Population density} = \dfrac{\text{Number of people}}{\text{Area}}$

Singapore has an area of 7698 km² and a population density of 728/km²
Calculate the number of people in Singapore.
Give your answer to the nearest thousand.

(Total 2 marks)

18 The formula $C = \dfrac{D}{V}$ gives the fuel consumption, C km/l, of a car which travels D km on V litres of fuel.
The fuel consumption of a *Hornet* is 16.8 km/l and it travels 756 km on a full tank of fuel.
Calculate the number of litres of fuel a *Hornet's* tank holds when it is full.

(Total 2 marks)

19 The formula $E = \dfrac{100V}{D}$ gives the fuel economy, E litres/100 km, of a car which travels D km on V litres of fuel.

The fuel economy of a *Hawk* is 5.6 litres/100 km.

It travels 325 km.

Calculate the amount of fuel the *Hawk* uses.

(Total 2 marks)

20 $\text{Power} = \dfrac{\text{Energy consumed}}{\text{Time}}$

where Power is in Watts, Energy is in Joules and Time is in seconds.

The power rating of a hairdryer is 2.2 kiloWatts.

Esther uses the hairdryer for 3 minutes 15 seconds.

Calculate the energy, in Joules, it consumes.

(Total 3 marks)

23 Construction

1 The length of each side of a rhombus is 4.6 cm.
The length of its longer diagonal is 8.4 cm.
Use ruler and compasses to construct an accurate, full-size drawing of the rhombus.
You must show all construction lines.

(Total 3 marks)

2 In triangle ABC, $AB = 6.1$ cm, $AC = 6.9$ cm and $BC = 5.4$ cm.
Use ruler and compasses to construct:
a) an accurate, full-size drawing of the triangle
b) the bisector of angle BAC.
You must show all construction lines.

(Total 4 marks)

3 In triangle PQR, $PQ = 5.6$ cm, $PR = 6.3$ cm and angle $QPR = 58°$.
a) Make an accurate full-size drawing of triangle PQR.   **(2)**
b) Use ruler and compasses to construct the perpendicular bisector of the
side QR.
You must show all construction lines.   **(2)**

(Total 4 marks)

4 The lengths of two of the sides of a parallelogram are 6.7 cm and 4.6 cm.
The length of its shorter diagonal is 5.1 cm.
a) Use ruler and compasses to construct an accurate, full-size drawing of the
parallelogram.   **(3)**
b) Use ruler and compasses to construct the perpendicular bisector of the longer
diagonal.
You must show all construction lines.   **(2)**

(Total 5 marks)

5 A ladder is 4.8 m long. The ladder leans against a vertical wall with the foot of the
ladder resting on horizontal ground.
The ladder reaches up the wall a distance of 4.3 m.
Using a scale of 1 : 50, make an accurate scale drawing and from it find:
a) the distance of the ladder from the bottom of the wall
b) the angle the ladder makes with the ground.

(Total 3 marks)

6

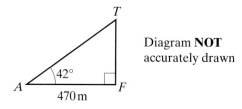

Diagram **NOT** accurately drawn

The Menara tower is in Kuala Lumpur, Malaysia.
F is the foot of the tower and T is the top of the tower.
A is a point such that AF is horizontal and $AF = 470$ m.
From A, the angle of elevation of T is 42°.
Using a scale of 1 cm to 50 m, make an accurate scale drawing and from it work out the height of the tower.

(Total 3 marks)

7

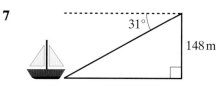

Diagram **NOT** accurately drawn

From the top of a cliff, the angle of depression of a boat is 31°.
The height of the cliff is 148 m.
Using a scale of 1 cm to 20 m, make an accurate scale drawing and from it work out the distance from the boat to the foot of the cliff.

(Total 3 marks)

8   A ship sails 126 km on a bearing of 058° from a port P to a point A.
From A, the ship sails 154 km on a bearing of 134° to a point B.
Using a scale of 1 cm to 20 km, make an accurate scale drawing and from it find:
a)   the distance PB **(2)**
b)   the bearing on which the ship must sail to return directly from B to P.   **(2)**

(Total 4 marks)

24 Geometry

1

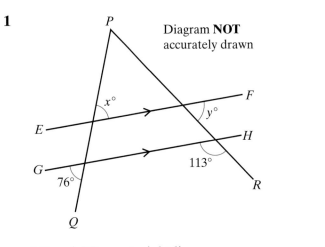

Diagram **NOT** accurately drawn

PQ and *PR* are straight lines.
EF and *GH* are parallel lines.
a) Find the value of *x*.
  Give a reason for each step in your working. **(2)**
b) Find the value of *y*.
  Give a reason for each step in your working. **(2)**

(Total 4 marks)

2

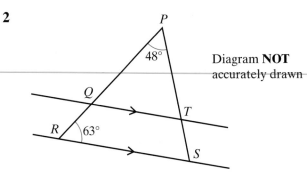

Diagram **NOT** accurately drawn

In the diagram, *PQR* and *PTS* are straight lines.
QT is parallel to *RS*.
Angle *PRS* = 63°
Angle *RPS* = 48°
Work out the size of angle *PTQ*.
Give a reason for each step in your working.

(Total 3 marks)

3

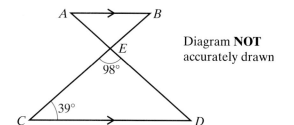

Diagram **NOT**
accurately drawn

AB is parallel to *CD*.
AD and *BC* are straight lines which intersect at *E*.
Angle *DCE* = 39°
Angle *CED* = 98°
Work out the size of angle *BAE*.
Give a reason for each step in your working.

(Total 3 marks)

4

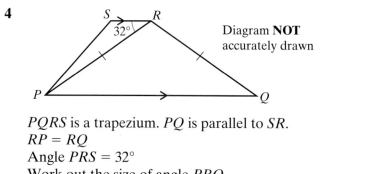

Diagram **NOT**
accurately drawn

PQRS is a trapezium. *PQ* is parallel to *SR*.
RP = *RQ*
Angle *PRS* = 32°
Work out the size of angle *PRQ*.
Give a reason for each step in your working.

(Total 5 marks)

5 **a)** Work out the sum of the interior angles of an octagon. **(2)**
 b) Work out the size of each exterior angle of a **regular** 10-sided polygon. **(2)**

(Total 4 marks)

6 **a)** Work out the size of each interior angle of a regular 9-sided polygon. **(2)**
 b)

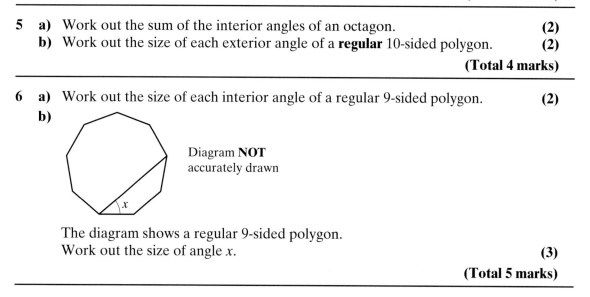

Diagram **NOT**
accurately drawn

The diagram shows a regular 9-sided polygon.
Work out the size of angle *x*. **(3)**

(Total 5 marks)

7 **a)** The sum of the interior angles of a polygon is $1800°$.
Work out how many sides the polygon has. **(2)**
b) The sizes of four of the exterior angles of a pentagon are $44°$, $81°$, $109°$ and $62°$.
Work out the size of the other exterior angle. **(2)**

(Total 4 marks)

---

8 The size of each exterior angle of a regular polygon is $20°$.
a) Work out the number of sides the polygon has. **(2)**
b) Work out the sum of the interior angles of the polygon. **(2)**

(Total 4 marks)

---

9 The size of each interior angle of a regular polygon is $156°$.
Work out how many sides the polygon has.

(Total 3 marks)

10

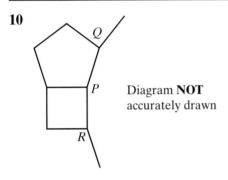

Diagram **NOT**
accurately drawn

Two sides of a regular pentagon, a square and a regular n-sided polygon meet at the point P.
PQ and PR are adjacent sides of the regular n-sided polygon.
Work out the value of n.

(Total 4 marks)

11

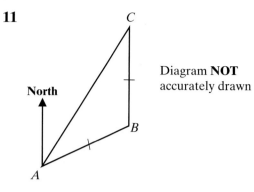

Diagram **NOT** accurately drawn

The bearing of *B* from *A* is 076°
C is due North of *B*.
AB = BC

a) Work out the bearing of *C* from *A*. (2)
b) Work out the bearing of *A* from *C*. (2)

(Total 4 marks)

12

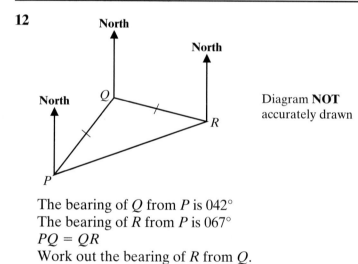

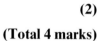

Diagram **NOT** accurately drawn

The bearing of *Q* from *P* is 042°
The bearing of *R* from *P* is 067°
PQ = QR
Work out the bearing of *R* from *Q*.

(Total 4 marks)

25 Transformations

1

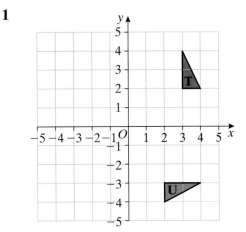

  a) Describe fully the single transformation which maps triangle **T** onto triangle **U**. **(3)**

  b) On a copy of the grid, reflect triangle **T** in the line $x = 1$

       Label your new triangle **V**. **(2)**

       (Total 5 marks)

2

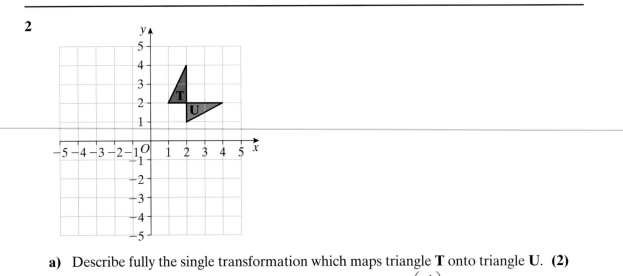

  a) Describe fully the single transformation which maps triangle **T** onto triangle **U**. **(2)**

  b) On a copy of the grid, translate triangle **T** by the vector $\begin{pmatrix} 1 \\ -6 \end{pmatrix}$

       Label your new triangle **V**. **(2)**

       (Total 4 marks)

3

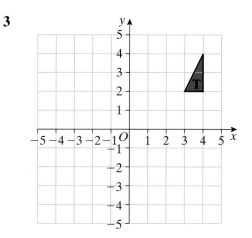

a) On a copy of the grid, rotate triangle **T** through 180° about the point (1, 0).
Label your new triangle **U**. **(3)**

b) On a copy of the grid, reflect triangle **T** in the line $y = 3$
Label your new triangle **V**. **(2)**

(Total 5 marks)

4

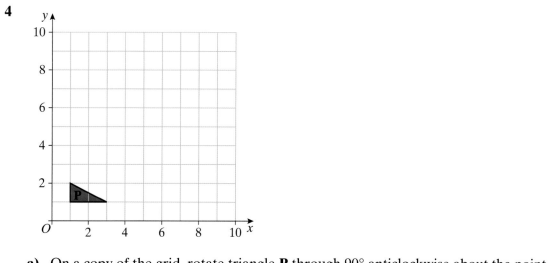

a) On a copy of the grid, rotate triangle **P** through 90° anticlockwise about the point (2, 7).
Label your new triangle **Q**. **(2)**

b) On a copy of the grid, enlarge triangle **P** with scale factor 3 and centre (0, 0).
Label your new triangle **R**. **(2)**

(Total 4 marks)

5

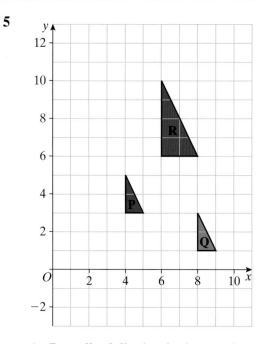

a) Describe fully the single transformation which maps triangle **P** onto triangle **Q**. **(2)**

b) Describe fully the single transformation which maps:
   i) triangle **P** onto triangle **R**
   ii) triangle **R** onto triangle **P**. **(3)**

(Total 5 marks)

6

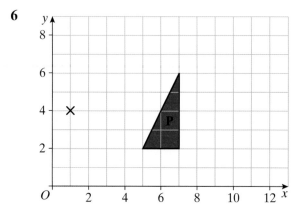

On a copy of the grid, enlarge triangle **P** with scale factor $1\frac{1}{2}$ and centre $(1, 4)$

(Total 3 marks)

7

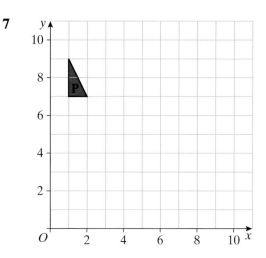

Reflect triangle **P** in the line $y = 6$ to give triangle **Q**.
Reflect triangle **Q** in the line $x = 5$ to give triangle **R**.
Describe fully the single transformation which maps triangle **P** onto triangle **R**.

(Total 4 marks)

8

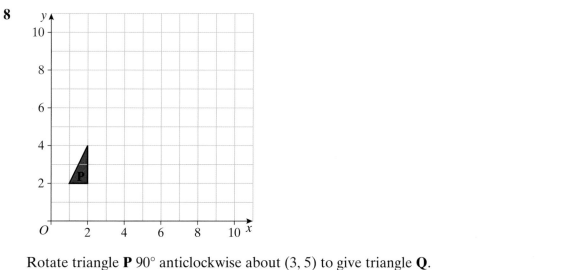

Rotate triangle **P** 90° anticlockwise about (3, 5) to give triangle **Q**.
Rotate triangle **Q** 90° clockwise about (7, 6) to give triangle **R**.
Describe fully the single transformation which maps triangle **P** onto triangle **R**.

(Total 4 marks)

26 Circle properties

1

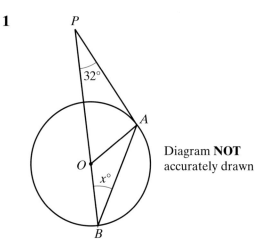

Diagram **NOT** accurately drawn

PA is a tangent at A to a circle, centre O.
POB is a straight line.
B is a point on the circle.
Angle $APB = 32°$
Find the value of x.

(Total 3 marks)

2

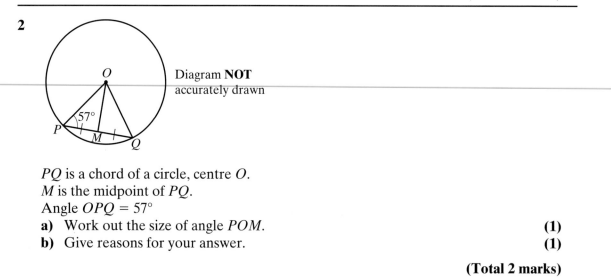

Diagram **NOT** accurately drawn

PQ is a chord of a circle, centre O.
M is the midpoint of PQ.
Angle $OPQ = 57°$
a) Work out the size of angle POM. **(1)**
b) Give reasons for your answer. **(1)**

(Total 2 marks)

3

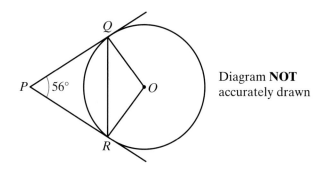

Diagram **NOT** accurately drawn

PQ and *PR* are tangents to a circle with centre *O*.
Angle *QPR* = 56°
Calculate the size of angle *ORQ*.   **(Total 2 marks)**

4

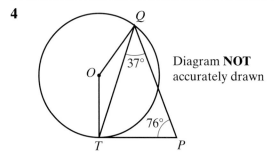

Diagram **NOT** accurately drawn

PT is a tangent at *T* to a circle, centre *O*.
Q is a point on the circle.
Angle *PQT* = 37° and angle *QPT* = 76°
Work out the size of angle *QOT*.   **(Total 2 marks)**

5

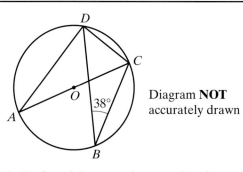

Diagram **NOT** accurately drawn

A, *B*, *C* and *D* are points on the circumference of a circle, centre *O*.
AC is a diameter of the circle.
Angle *CBD* = 38°
a) i) Find the size of angle *DAC*.
   ii) Give a reason for your answer.   **(2)**
b) Find the size of angle *ACD*.   **(2)**
                    (Total 4 marks)

6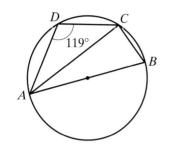

Diagram **NOT**
accurately drawn

A, B, C and D are points on a circle.
AB is a diameter of the circle. Angle $ADC = 119°$
a) i) Work out the size of angle ABC.
   ii) Give a reason for your answer.   **(2)**
b) Work out the size of angle BAC.   **(2)**

(Total 4 marks)

7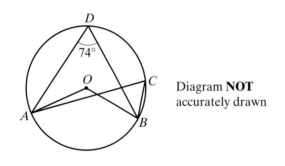

Diagram **NOT**
accurately drawn

A, B, C and D are points on the circumference of a circle, centre O.
Angle $ADB = 74°$
a) i) Find the size of angle AOB.
   ii) Give a reason for your answer.   **(2)**
b) i) Find the size of angle ACB.
   ii) Give a reason for your answer.   **(2)**

(Total 4 marks)

8

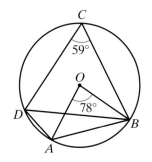

Diagram **NOT** accurately drawn

A, B, C and D are points on a circle with centre O.
Angle $AOB = 78°$ and angle $BCD = 59°$
a) i) Calculate the size of angle ADB.
   ii) Give a reason for your answer. **(2)**
b) Calculate the size of angle DBO.
   Give a reason for each step in your working. **(3)**

(Total 5 marks)

9

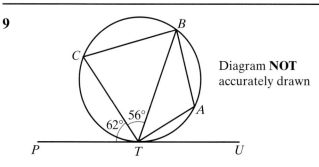

Diagram **NOT** accurately drawn

A, B, C and T are points on a circle.
PU is a tangent to the circle at T.
Angle $PTC = 62°$ and angle $BTC = 56°$
a) i) Find the size of angle CBT.
   ii) Give a reason for your answer. **(2)**
b) Find the size of angle BAT. **(2)**
c) Explain why CB is parallel to PU. **(1)**

(Total 5 marks)

10

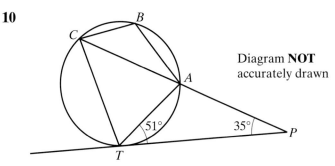

Diagram **NOT**
accurately drawn

A, B, C and T are points on a circle.
PT is a tangent to the circle at T.
Angle $ATP = 51°$ and angle $APT = 35°$
a) i) Find the size of angle ACT.
   ii) Give a reason for your answer. **(2)**
b) Find the size of angle ABC. **(3)**

(Total 5 marks)

11

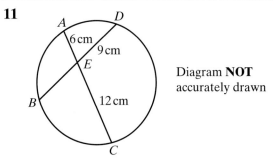

Diagram **NOT**
accurately drawn

A, B, C and D are points on a circle.
The chords AC and BD intersect at E.
$AE = 6$ cm, $CE = 12$ cm and $DE = 9$ cm
Calculate the length of BE.

(Total 3 marks)

12

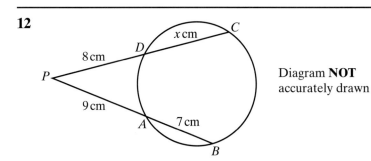

Diagram **NOT**
accurately drawn

A, B, C and D are points on a circle.
PAB and PDC are straight lines.
$PA = 9$ cm, $AB = 7$ cm and $PD = 8$ cm
Calculate the value of x.

(Total 3 marks)

27 Area and perimeter

1

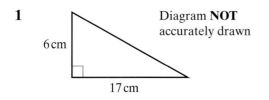

Diagram **NOT**
accurately drawn

Work out the area of the triangle.

(Total 2 marks)

2

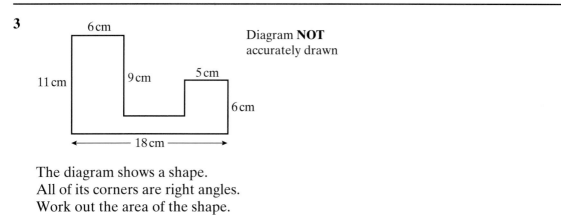

Diagram **NOT**
accurately drawn

The area of the triangle is 56 cm².
Work out the value of h.

(Total 2 marks)

3

Diagram **NOT**
accurately drawn

The diagram shows a shape.
All of its corners are right angles.
Work out the area of the shape.

(Total 4 marks)

4

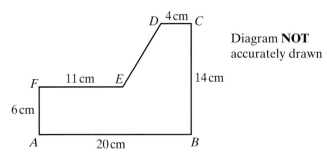

The diagram shows a shape $ABCDEF$.
The corners at A, B, C and F are right angles.
Work out the area of the shape.

(Total 4 marks)

5

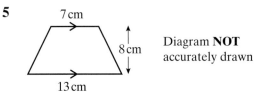

The lengths of the parallel sides of a trapezium are 13 cm and 7 cm.
The distance between the parallel sides of the trapezium is 8 cm.
Work out the area of the trapezium.

(Total 2 marks)

6

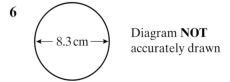

Diagram **NOT**
accurately drawn

A circle has a diameter of 8.3 cm.
Work out the circumference of the circle.
Give your answer correct to 3 significant figures.

(Total 2 marks)

7

Diagram **NOT**
accurately drawn

A circle has a radius of 5.3 cm.
Work out the area of the circle.
Give your answer correct to 3 significant figures.

(Total 2 marks)

8 The circumference of a circle is 23.7 cm.
Work out the radius of the circle.
Give your answer correct to 3 significant figures.

(Total 2 marks)

9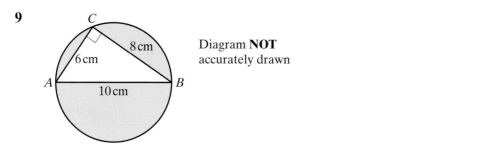

The diagram shows a right-angled triangle ABC and a circle.
A, B and C are points on the circle.
AB is the diameter of the circle.
$AB = 10$ cm, $AC = 6$ cm, $BC = 8$ cm.
Work out the area of the shaded part of the circle.
Give your answer correct to 3 significant figures.

(Total 5 marks)

10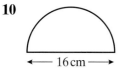

Diagram **NOT** accurately drawn

The diameter of a semicircle is 16 cm.
Work out the perimeter of the semicircle.
Give your answer correct to 3 significant figures.

(Total 4 marks)

11

Diagram **NOT** accurately drawn

A quarter circle has a radius of 6.5 cm.
Work out the perimeter of the quarter circle.
Give your answer correct to 3 significant figures.

(Total 4 marks)

12

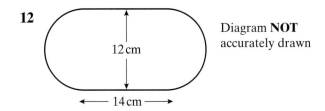

Diagram **NOT**
accurately drawn

A shape is made from a rectangle and two semicircles.
The length of the rectangle is 14 cm and its width is 12 cm.
Work out the area of the shape.
Give your answer correct to 3 significant figures.

(Total 4 marks)

13

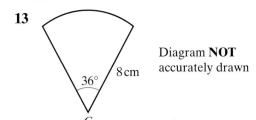

Diagram **NOT**
accurately drawn

The diagram shows the sector of a circle, centre C.
The radius of the circle is 8 cm.
The angle at the centre of the circle is 36°
Calculate the area of the sector.
Give your answer correct to 3 significant figures.

(Total 3 marks)

14

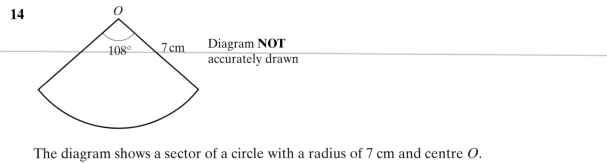

Diagram **NOT**
accurately drawn

The diagram shows a sector of a circle with a radius of 7 cm and centre O.
The angle at the centre of the circle is 108°
Calculate the perimeter of this sector.
Give your answer correct to 3 significant figures.

(Total 4 marks)

15 Change 3.7 m² to cm².
Give your answer correct to 3 significant figures.

(Total 2 marks)

28 3-D shapes – volume and surface area

1

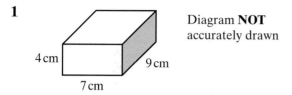

Diagram **NOT**
accurately drawn

The diagram shows a cuboid.
Work out the total surface area of the cuboid.   **(Total 3 marks)**

2

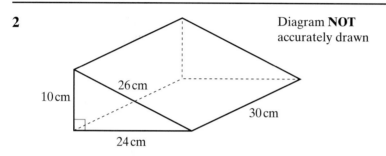

Diagram **NOT**
accurately drawn

The diagram shows a prism.
The cross-section of the prism is a right-angled triangle.
The lengths of the sides of the triangle are 10 cm, 24 cm and 26 cm.
The length of the prism is 30 cm.
a) Work out the volume of the prism.   **(3)**
b) Work out the total surface area of the prism.   **(3)**

(Total 6 marks)

3

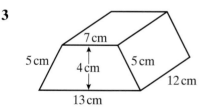

Diagram **NOT**
accurately drawn

The diagram shows a prism.
The cross-section of the prism is a trapezium.
The lengths of the parallel sides of the trapezium are 13 cm and 7 cm.
The distance between the parallel sides of the trapezium is 4 cm.
The length of each of the other two sides of the trapezium is 5 cm.
The length of the prism is 12 cm.
a) Work out the volume of the prism.   **(3)**
b) Work out the total surface area of the prism.   **(3)**

(Total 6 marks)

4

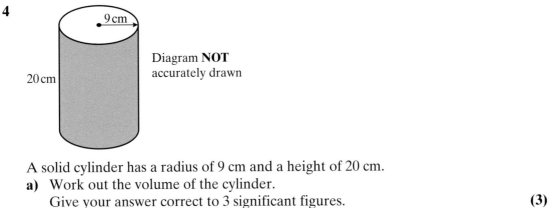

Diagram **NOT** accurately drawn

A solid cylinder has a radius of 9 cm and a height of 20 cm.
a) Work out the volume of the cylinder.
Give your answer correct to 3 significant figures. **(3)**
b) Work out the total surface area of the cylinder.
Give your answer correct to 3 significant figures. **(3)**
(Total 6 marks)

5

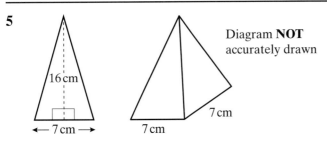

Diagram **NOT** accurately drawn

A solid pyramid has a square base of side 7 cm.
The vertical height of each of its triangular faces is 16 cm.
Work out the total surface area of the pyramid.

(Total 3 marks)

6

Diagram **NOT** accurately drawn

The diagram shows a solid sphere with a radius of 5.9 cm.
a) Calculate the surface area of the sphere.
Give your answer correct to 3 significant figures. **(2)**
b) Calculate the volume of the sphere.
Give your answer correct to 3 significant figures. **(2)**
(Total 4 marks)

7

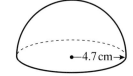

Diagram **NOT** accurately drawn

A solid hemisphere has a radius of 4.7 cm.
Calculate the **total** surface area of the hemisphere.
Give your answer correct to 3 significant figures.

(Total 3 marks)

8

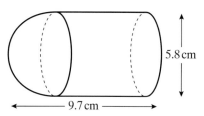

5.8 cm

Diagram **NOT** accurately drawn

9.7 cm

The diagram shows a solid shape made from a cylinder and a hemisphere.
Both the cylinder and the hemisphere have a diameter of 5.8 cm.
The length of the solid is 9.7 cm.
Calculate the volume of the solid.
Give your answer correct to 3 significant figures.

(Total 3 marks)

9

30 cm

26 cm

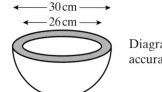

Diagram **NOT** accurately drawn

The diagram shows a hemispherical wooden bowl.
The bowl has an external diameter of 30 cm and an internal diameter of 26 cm.
Calculate the volume of wood.
Give your answer correct to 3 significant figures.

(Total 3 marks)

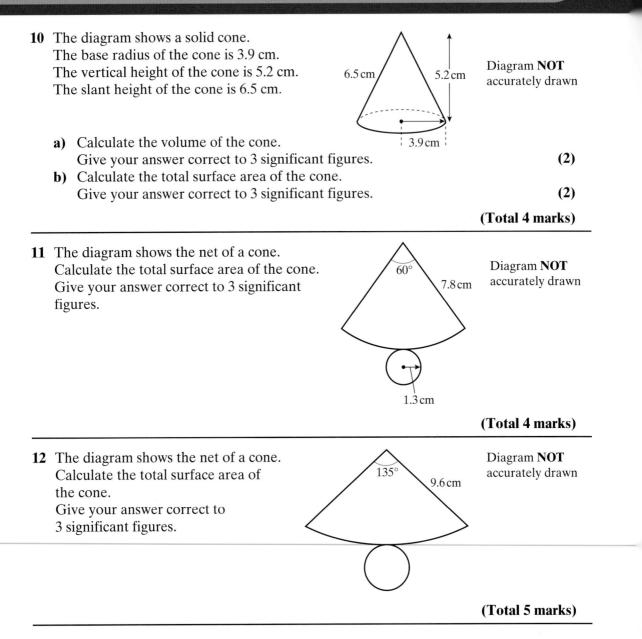

10 The diagram shows a solid cone.
The base radius of the cone is 3.9 cm.
The vertical height of the cone is 5.2 cm.
The slant height of the cone is 6.5 cm.

6.5 cm   5.2 cm

Diagram **NOT**
accurately drawn

3.9 cm

a) Calculate the volume of the cone.
Give your answer correct to 3 significant figures. **(2)**

b) Calculate the total surface area of the cone.
Give your answer correct to 3 significant figures. **(2)**

(Total 4 marks)

11 The diagram shows the net of a cone.
Calculate the total surface area of the cone.
Give your answer correct to 3 significant
figures.

60°   7.8 cm

Diagram **NOT**
accurately drawn

1.3 cm

(Total 4 marks)

12 The diagram shows the net of a cone.
Calculate the total surface area of
the cone.
Give your answer correct to
3 significant figures.

135°   9.6 cm

Diagram **NOT**
accurately drawn

(Total 5 marks)

13

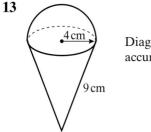

Diagram **NOT**
accurately drawn

The diagram shows a solid shape made from a hemisphere and a cone.
The hemisphere has a radius of 4 cm.
The cone has a radius of 4 cm and a slant height of 9 cm.
Calculate the total surface area of the solid.
Give your answer correct to 3 significant figures.

(Total 3 marks)

14

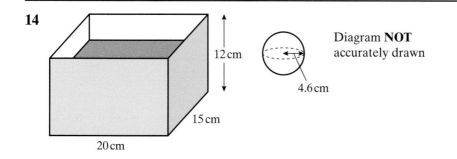

Diagram **NOT**
accurately drawn

A rectangular container is 20 cm long, 15 cm wide and 12 cm high.
The container is filled with water to a depth of 10 cm.
A metal sphere of radius 4.6 cm is dropped into the water and sinks to the bottom.
Calculate the rise in the water level.
Give your answer correct to 3 significant figures.

(Total 5 marks)

15 Change 0.64 m^3 to cm^3.

(Total 2 marks)

29 Pythagoras' theorem

1

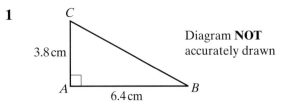

Diagram **NOT** accurately drawn

In triangle ABC, angle $A = 90°$, $AB = 6.4$ cm and $AC = 3.8$ cm.
Work out the length of BC.
Give your answer correct to 3 significant figures.

(Total 3 marks)

2

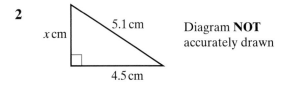

Diagram **NOT** accurately drawn

Work out the value of x.

(Total 3 marks)

3

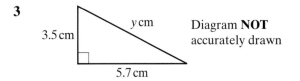

Diagram **NOT** accurately drawn

Calculate the value of y.
Give your answer correct to 3 significant figures.

(Total 3 marks)

4

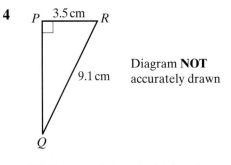

Diagram **NOT** accurately drawn

PQR is a right-angled triangle.
$PR = 3.5$ cm and $QR = 9.1$ cm.
Calculate the length of PQ.

(Total 3 marks)

5

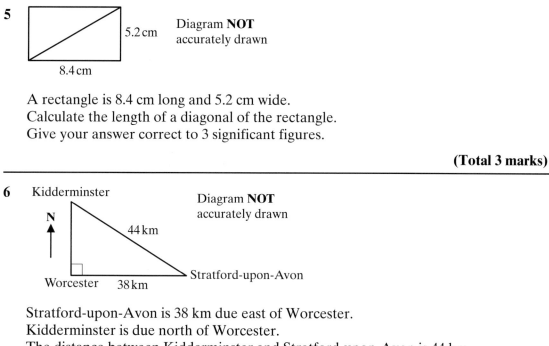

A rectangle is 8.4 cm long and 5.2 cm wide.
Calculate the length of a diagonal of the rectangle.
Give your answer correct to 3 significant figures.

(Total 3 marks)

6 Kidderminster

Stratford-upon-Avon is 38 km due east of Worcester.
Kidderminster is due north of Worcester.
The distance between Kidderminster and Stratford-upon-Avon is 44 km.
Calculate the distance between Kidderminster and Worcester.
Give your answer in kilometres, correct to 1 decimal place.

(Total 3 marks)

7

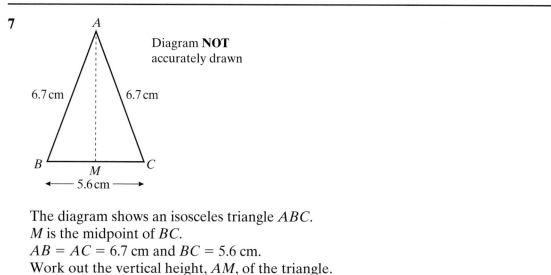

The diagram shows an isosceles triangle ABC.
M is the midpoint of BC.
$AB = AC = 6.7$ cm and $BC = 5.6$ cm.
Work out the vertical height, AM, of the triangle.
Give your answer correct to 3 significant figures.

(Total 3 marks)

8

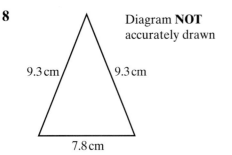

Diagram **NOT** accurately drawn

9.3 cm 9.3 cm

7.8 cm

The diagram shows an isosceles triangle.
Calculate the area of the triangle.
Give your answer correct to 3 significant figures.

(Total 4 marks)

9

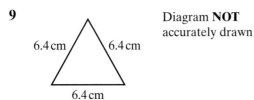

Diagram **NOT** accurately drawn

6.4 cm 6.4 cm

6.4 cm

The diagram shows an equilateral triangle of side 6.4 cm.
Calculate the area of the triangle.
Give your answer correct to 3 significant figures.

(Total 3 marks)

10 Two points, A and B, are plotted on a centimetre grid.
A has coordinates $(3, 1)$ and B has coordinates $(7, 8)$.
Calculate the length of AB.
Give your answer correct to 3 significant figures.

(Total 4 marks)

11 P is the point with coordinates $(-4, 3)$.
Q is the point with coordinates $(8, -2)$.
Work out the length of PQ.

(Total 4 marks)

12

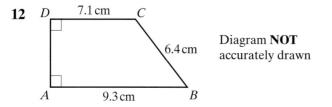

Diagram **NOT** accurately drawn

ABCD is a trapezium, in which *AB* is parallel to *DC*.
Angle *A* = 90° and angle *D* = 90°
AB = 9.3 cm. *BC* = 6.4 cm. *CD* = 7.1 cm.
Work out the area of trapezium *ABCD*.
Give your answer correct to 3 significant figures.

(Total 5 marks)

13

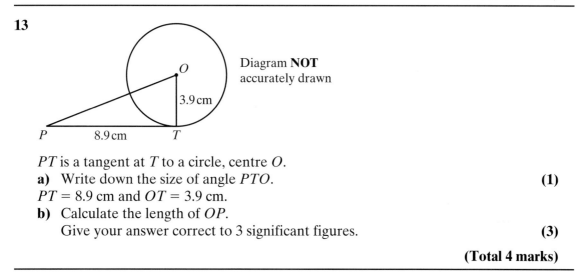

Diagram **NOT** accurately drawn

PT is a tangent at *T* to a circle, centre *O*.
a) Write down the size of angle *PTO*. **(1)**
PT = 8.9 cm and *OT* = 3.9 cm.
b) Calculate the length of *OP*.
Give your answer correct to 3 significant figures. **(3)**

(Total 4 marks)

14

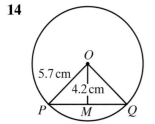

Diagram **NOT** accurately drawn

PQ is a chord of a circle, centre *O*.
M is the midpoint of *PQ*.
The radius of the circle is 5.7 cm.
OM = 4.2 cm.
Work out the length of the chord *PQ*.
Give your answer correct to 3 significant figures.

(Total 4 marks)

15

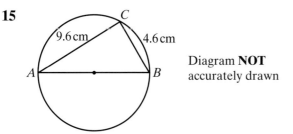

Diagram **NOT** accurately drawn

A, B and C are points on a circle.
AB is a diameter of the circle.
$AC = 9.6$ cm and $BC = 4.6$ cm.
Calculate the diameter of the circle.
Give your answer correct to 3 significant figures.

(Total 3 marks)

16

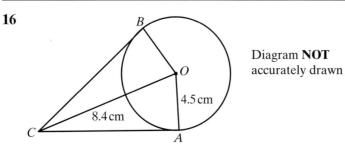

Diagram **NOT** accurately drawn

A and B are points on a circle, centre O.
The lines CA and CB are tangents to the circle.
$OA = 4.5$ cm and $OC = 8.4$ cm.
Calculate the perimeter of the kite $CAOB$.
Give your answer correct to 3 significant figures.

(Total 5 marks)

17

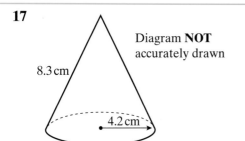

Diagram **NOT** accurately drawn

The diagram shows a solid cone.
The base radius of the cone is 4.2 cm.
The slant height of the cone is 8.3 cm.
Calculate the volume of the cone.
Give your answer correct to 3 significant figures.

(Total 4 marks)

18

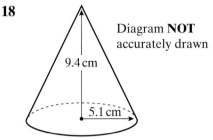

Diagram **NOT** accurately drawn

9.4 cm

5.1 cm

The diagram shows a solid cone.
The base radius of the cone is 5.1 cm.
The vertical height of the cone is 9.4 cm.
Calculate the curved surface area of the cone.
Give your answer correct to 3 significant figures.

(Total 4 marks)

19

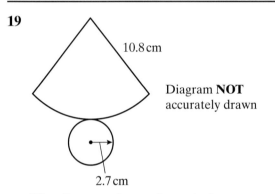

10.8 cm

Diagram **NOT** accurately drawn

2.7 cm

The diagram shows the net of a cone.
Calculate the volume of the cone.
Give your answer correct to 3 significant figures.

(Total 4 marks)

20

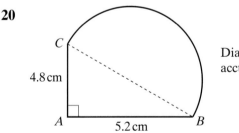

Diagram **NOT** accurately drawn

C

4.8 cm

A 5.2 cm B

A shape is made from a right-angled triangle, ABC, and a semicircle.
In triangle ABC, angle $A = 90°$, $AB = 5.2$ cm and $AC = 4.8$ cm.
BC is the diameter of the semicircle.
Calculate the area of the shape.
Give your answer correct to 3 significant figures.

(Total 5 marks)

21

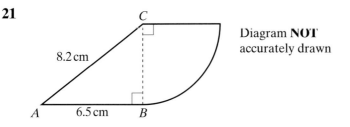

Diagram **NOT** accurately drawn

A shape is made from a right-angled triangle, *ABC*, and a quarter circle.
In triangle *ABC*, angle $B = 90°$, $AB = 6.5$ cm and $AC = 8.2$ cm.
C is the centre of the quarter circle.
Calculate the perimeter of the shape.
Give your answer correct to 3 significant figures.

(Total 5 marks)

22

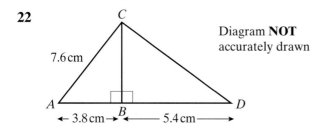

Diagram **NOT** accurately drawn

Angle $ABC = 90°$ and angle $DBC = 90°$
$AB = 3.8$ cm, $AC = 7.6$ cm and $BD = 5.4$ cm.
Calculate the length of *CD*.
Give your answer correct to 3 significant figures.

(Total 4 marks)

23

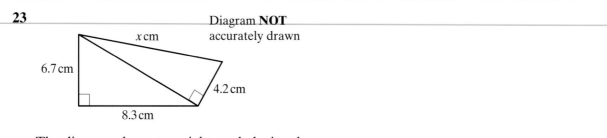

Diagram **NOT** accurately drawn

The diagram shows two right-angled triangles.
Work out the value of *x*.
Give your answer correct to 3 significant figures.

(Total 4 marks)

24

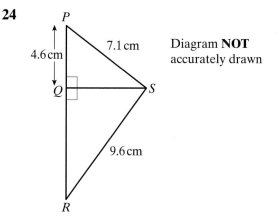

Diagram **NOT** accurately drawn

Angle $PQS = 90°$ and angle $RQS = 90°$
$PQ = 4.6$ cm, $PS = 7.1$ cm and $RS = 9.6$ cm.
Calculate the length of QR.
Give your answer correct to 3 significant figures.

(Total 4 marks)

25

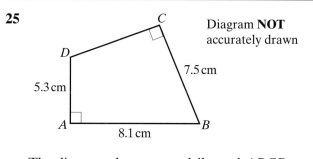

Diagram **NOT** accurately drawn

The diagram shows a quadrilateral $ABCD$.
Angle $A = 90°$ and angle $C = 90°$
$AB = 8.1$ cm, $AD = 5.3$ cm and $BC = 7.5$ cm.
Calculate the area of quadrilateral $ABCD$.
Give your answer correct to 3 significant figures.

(Total 5 marks)

30 Trigonometry

1

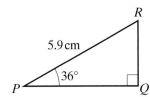

Diagram **NOT** accurately drawn

The diagram shows triangle PQR.
Angle $PQR = 90°$, $PR = 5.9$ cm and angle $RPQ = 36°$
Work out the length of QR.
Give your answer correct to 3 significant figures.

(Total 3 marks)

2

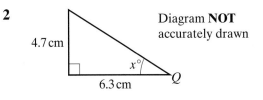

Diagram **NOT** accurately drawn

Calculate the value of x.
Give your answer correct to 1 decimal place.

(Total 3 marks)

3

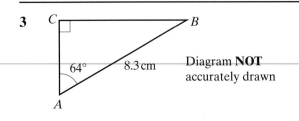

Diagram **NOT** accurately drawn

Triangle ABC is right-angled at C.
$AB = 8.3$ cm and angle $BAC = 64°$
Calculate the length of AC.
Give your answer correct to 3 significant figures.

(Total 3 marks)

4

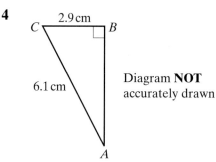

Diagram **NOT**
accurately drawn

ABC is a triangle.
Angle $ABC = 90°$, $AC = 6.1$ cm and $BC = 2.9$ cm.
Work out the size of angle BCA.
Give your answer correct to 1 decimal place.

(Total 3 marks)

5

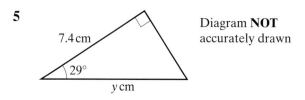

Diagram **NOT**
accurately drawn

Work out the value of y.
Give your answer correct to 3 significant figures.

(Total 3 marks)

6

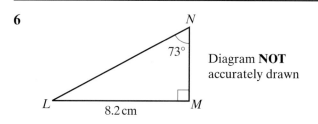

Diagram **NOT**
accurately drawn

LMN is a right-angled triangle.
$LM = 8.2$ cm, angle $LMN = 90°$ and angle $LNM = 73°$
Calculate the length of MN.
Give your answer correct to 3 significant figures.

(Total 3 marks)

7

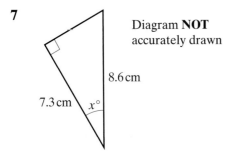

Diagram **NOT** accurately drawn

8.6 cm

7.3 cm $x°$

Calculate the value of x.
Give your answer correct to 1 decimal place.

(Total 3 marks)

8

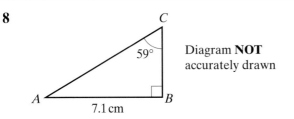

Diagram **NOT** accurately drawn

Triangle ABC is right-angled at B.
$AB = 7.1$ cm and angle $C = 59°$
Work out the length of AC.
Give your answer correct to 3 significant figures.

(Total 3 marks)

9

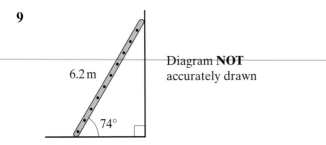

Diagram **NOT** accurately drawn

A ladder is 6.2 m long.
The ladder rests against a vertical wall.
The foot of the ladder rests on horizontal ground.
The ladder makes an angle of 74° with the ground.
Work out how far the foot of the ladder is from the bottom of the wall.
Give your answer correct to 3 significant figures.

(Total 3 marks)

10

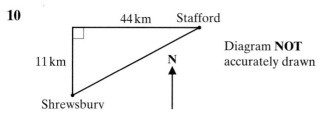

44 km   Stafford

Diagram **NOT**
accurately drawn

11 km

N

Shrewsbury

Stafford is 11 km north of Shrewsbury and 44 km east of Shrewsbury.
Calculate the bearing of Stafford from Shrewsbury.
Give your answer correct to the nearest degree.

(Total 3 marks)

11

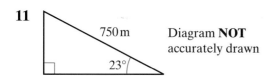

750 m   Diagram **NOT**
accurately drawn

23°

An aeroplane climbs for 750 m at an angle of 23° above the horizontal.
Calculate its increase in height.
Give your answer correct to 3 significant figures.

(Total 3 marks)

12 T

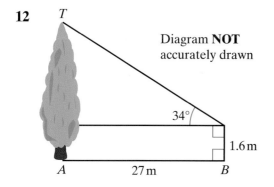

Diagram **NOT**
accurately drawn

34°

1.6 m

A 27 m B

Janine is standing on level ground at B, 27 metres away from the foot A of a tree TA.
She measures the angle of elevation of the top of the tree, at a height of 1.6 metres
above the ground, as 34°.
Calculate the height, TA, of the tree.
Give your answer correct to 3 significant figures.

(Total 4 marks)

13

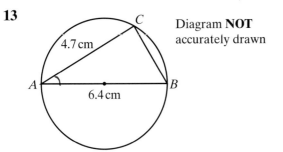

Diagram **NOT** accurately drawn

A, B and C are points on a circle.
AB is a diameter of the circle.
$AB = 6.4$ cm and $AC = 4.7$ cm.
Calculate the size of angle BAC.
Give your answer correct to 1 decimal place.

(Total 3 marks)

14

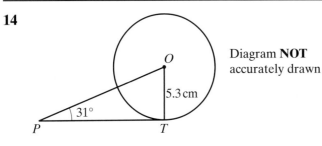

Diagram **NOT** accurately drawn

PT is a tangent at T to a circle, centre O.
$OT = 5.3$ cm
Angle $OPT = 31°$
Work out the length of OP.
Give your answer correct to 3 significant figures.

(Total 3 marks)

15

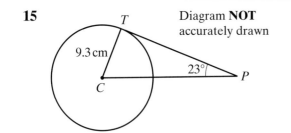

Diagram **NOT** accurately drawn

PT is a tangent at T to a circle, centre C.
$CT = 9.3$ cm
Angle $CPT = 23°$
Calculate the length of the tangent PT.
Give your answer correct to 3 significant figures.

(Total 4 marks)

16

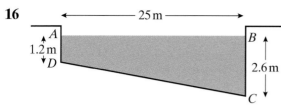

Diagram **NOT**
accurately drawn

The diagram shows the side view of a swimming pool.
The pool has vertical sides and a sloping bottom.
The water surface, AB, is horizontal and $AB = 25$ m.
The depth of water, AD, at the shallow end is 1.2 m.
The depth of water, BC, at the deep end is 2.6 m.
DC is the sloping bottom of the pool.
Calculate the size of the angle DC makes with the horizontal.
Give your answer correct to 1 decimal place.

(Total 4 marks)

17

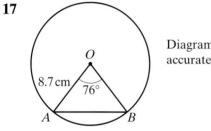

Diagram **NOT**
accurately drawn

AB is a chord of a circle, centre O.
The radius, OA, of the circle is 8.7 cm.
The chord AB subtends an angle of $76°$ at O.
Work out the length of AB.
Give your answer correct to 3 significant figures.

(Total 4 marks)

18

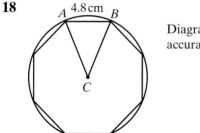

Diagram **NOT**
accurately drawn

The diagram shows a regular octagon inside a circle, centre C.
AB is a side of the octagon and $AB = 4.8$ cm.
Calculate the radius, AC, of the circle.
Give your answer correct to 3 significant figures.

(Total 5 marks)

19

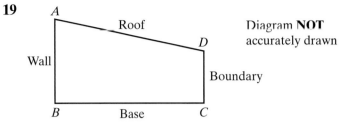

Diagram **NOT** accurately drawn

The trapezium *ABCD* shows one end of a carport.
The base *BC* is horizontal. *AB* and *CD* are vertical.
Carports are made with dimensions in the following ranges:

Wall height (*AB*)	1.95 m – 2.25 m
Base (*BC*)	1.80 m – 2.70 m
Boundary height (*CD*)	1.85 m – 2.10 m

Calculate the size of the greatest angle the roof can make with the horizontal.
Give your answer correct to 1 decimal place.

(Total 5 marks)

20

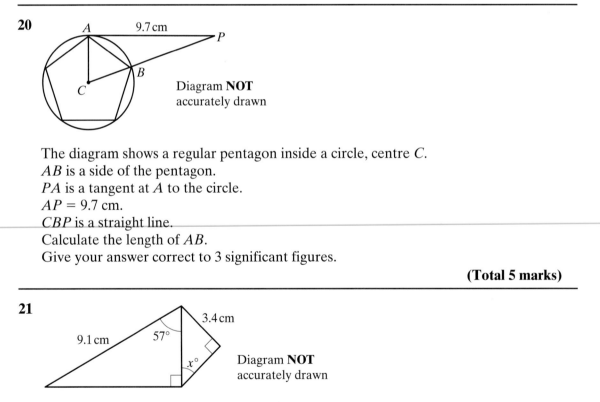

Diagram **NOT** accurately drawn

The diagram shows a regular pentagon inside a circle, centre *C*.
AB is a side of the pentagon.
PA is a tangent at *A* to the circle.
AP = 9.7 cm.
CBP is a straight line.
Calculate the length of *AB*.
Give your answer correct to 3 significant figures.

(Total 5 marks)

21

The diagram shows two right-angled triangles.
Calculate the value of *x*.
Give your answer correct to 1 decimal place.

Diagram **NOT** accurately drawn

(Total 4 marks)

22

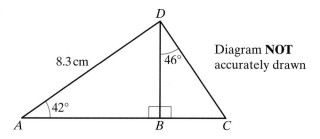

ABC is a straight line and angle $ABD = 90°$
$AD = 8.3$ cm, angle $BAD = 42°$ and angle $BDC = 46°$
Work out the length of CD.
Give your answer correct to 3 significant figures.

(Total 5 marks)

23

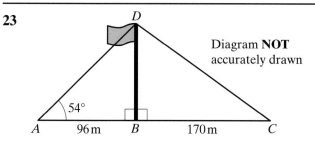

The diagram shows the Aqaba flagpole in Jordan, which is the tallest flagpole in the world.
The points A, B and C are in a straight line on horizontal ground.
The point B is at the foot of the flagpole and the point D is at the top of the flagpole.
$AB = 96$ m and $BC = 170$ m.
The angle of elevation of D from A is $54°$
Calculate the angle of elevation of D from C.
Give your answer correct to 1 decimal place.

(Total 4 marks)

24

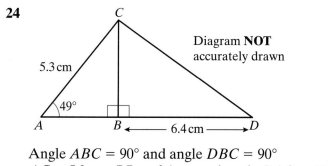

Angle $ABC = 90°$ and angle $DBC = 90°$
$AC = 5.3$ cm, $BD = 6.4$ cm and angle $BAC = 49°$
Calculate the length of CD.
Give your answer correct to 3 significant figures.

(Total 4 marks)

25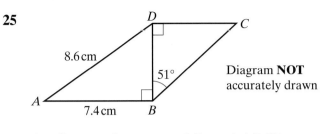

Diagram **NOT** accurately drawn

The diagram shows a quadrilateral $ABCD$.
$AB = 7.4$ cm and $AD = 8.6$ cm.
Angle $ABD = 90°$ and angle $BDC = 90°$
Angle $CBD = 51°$
Calculate the length of BC.
Give your answer correct to 3 significant figures.

(Total 5 marks)

31 Similar shapes

1 Triangles *ABC* and *DEF* are similar.

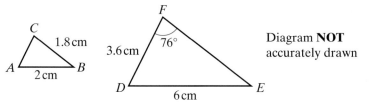

Diagram **NOT** accurately drawn

AB = 2 cm BC = 1.8 cm DE = 6 cm DF = 3.6 cm   Angle *DFE* = 76°

a) Find the size of angle *ACB*. **(1)**

b) Find the length of:

   i) *EF*

   ii) *AC*. **(4)**

(Total 5 marks)

2 Triangles *PQR* and *STU* are mathematically similar.

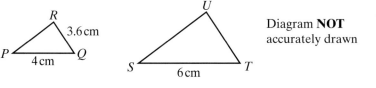

Diagram **NOT** accurately drawn

a) Work out the length of *TU*. **(2)**

The area of triangle *PQR* is 6 cm².

b) Work out the area of triangle *STU*. **(2)**

(Total 4 marks)

3

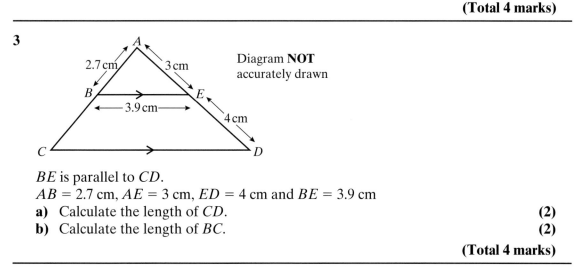

Diagram **NOT** accurately drawn

BE is parallel to *CD*.

AB = 2.7 cm, AE = 3 cm, ED = 4 cm and BE = 3.9 cm

a) Calculate the length of *CD*. **(2)**

b) Calculate the length of *BC*. **(2)**

(Total 4 marks)

4

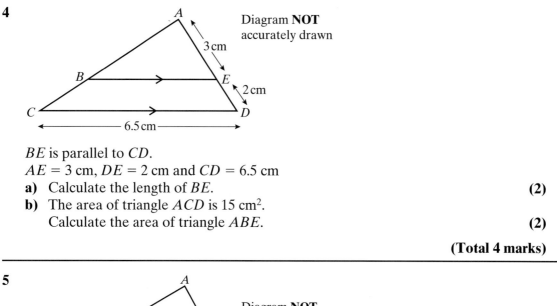

Diagram **NOT** accurately drawn

BE is parallel to *CD*.
AE = 3 cm, *DE* = 2 cm and *CD* = 6.5 cm
a) Calculate the length of *BE*. (2)
b) The area of triangle *ACD* is 15 cm².
    Calculate the area of triangle *ABE*. (2)

(Total 4 marks)

5

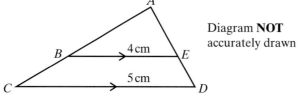

Diagram **NOT** accurately drawn

BE is parallel to *CD*.
BE = 4 cm and *CD* = 5 cm.
a) Find the ratio of the area of triangle *ACD* to the area of triangle *ABE*. (2)
b) Find the ratio of the area of triangle *ABE* to the area of trapezium *BCDE*. (1)

(Total 3 marks)

6

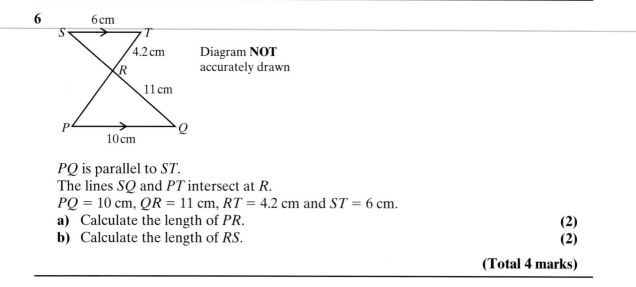

PQ is parallel to *ST*.
The lines *SQ* and *PT* intersect at *R*.
PQ = 10 cm, *QR* = 11 cm, *RT* = 4.2 cm and *ST* = 6 cm.
a) Calculate the length of *PR*. (2)
b) Calculate the length of *RS*. (2)

(Total 4 marks)

7

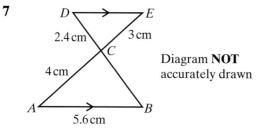

Diagram **NOT**
accurately drawn

AB is parallel to *DE*.
The lines *AE* and *BD* intersect at *C*.
AB = 5.6 cm, *AC* = 4 cm, *CD* = 2.4 cm and *CE* = 3 cm.
a) Calculate the length of *BC*. **(2)**
b) Calculate the length of *DE*. **(2)**
c) Find the ratio of the area of triangle *ABC* to the area of triangle *CDE*. **(1)**

(Total 5 marks)

8

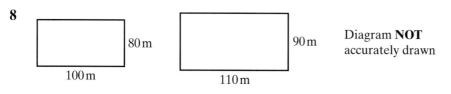

Diagram **NOT**
accurately drawn

One football pitch is a rectangle 100 m long and 80 m wide.
A second football pitch is a rectangle 110 m long and 90 m wide.
Are the two rectangles mathematically similar?
You must show working to justify your answer.

(Total 3 marks)

9

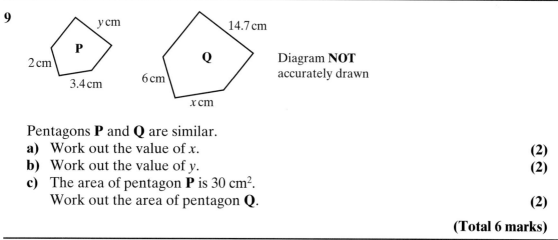

Diagram **NOT**
accurately drawn

Pentagons **P** and **Q** are similar.
a) Work out the value of *x*. **(2)**
b) Work out the value of *y*. **(2)**
c) The area of pentagon **P** is 30 cm².
Work out the area of pentagon **Q**. **(2)**

(Total 6 marks)

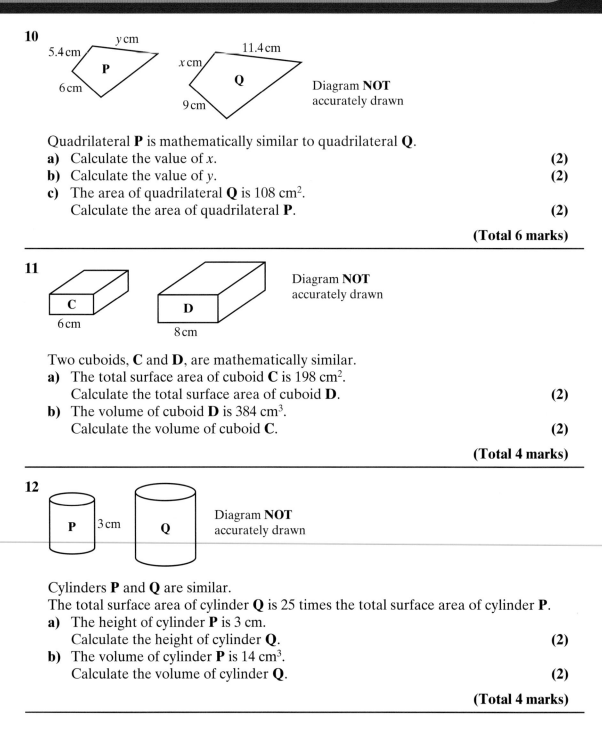

10

Quadrilateral **P** is mathematically similar to quadrilateral **Q**.
a) Calculate the value of *x*. (2)
b) Calculate the value of *y*. (2)
c) The area of quadrilateral **Q** is 108 cm².
     Calculate the area of quadrilateral **P**. (2)

(Total 6 marks)

11

Two cuboids, **C** and **D**, are mathematically similar.
a) The total surface area of cuboid **C** is 198 cm².
     Calculate the total surface area of cuboid **D**. (2)
b) The volume of cuboid **D** is 384 cm³.
     Calculate the volume of cuboid **C**. (2)

(Total 4 marks)

12

Cylinders **P** and **Q** are similar.
The total surface area of cylinder **Q** is 25 times the total surface area of cylinder **P**.
a) The height of cylinder **P** is 3 cm.
     Calculate the height of cylinder **Q**. (2)
b) The volume of cylinder **P** is 14 cm³.
     Calculate the volume of cylinder **Q**. (2)

(Total 4 marks)

13

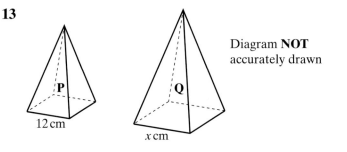

Diagram **NOT**
accurately drawn

12 cm

x cm

P and **Q** are square-based pyramids.
Pyramid **P** is mathematically similar to pyramid **Q**.
The volume of pyramid **P** is 384 cm³.
The volume of pyramid **Q** is 1296 cm³.

a) The base of pyramid **P** is a square of side 12 cm.
The base of pyramid **Q** is a square of side x cm.
Calculate the value of x. **(3)**

b) The total surface area of pyramid **P** is 444 cm².
Calculate the total surface area of pyramid **Q**. **(2)**

(Total 5 marks)

14

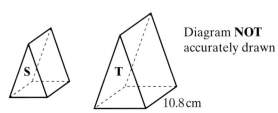

Diagram **NOT**
accurately drawn

10.8 cm

Prisms **S** and **T** are mathematically similar.
The total surface area of prism **S** is 175 cm².
The total surface area of prism **T** is 252 cm².

a) The length of prism **T** is 10.8 cm.
Calculate the length of prism **S**. **(3)**

b) The volume of prism **S** is 125 cm³.
Calculate the volume of prism **T**. **(2)**

(Total 5 marks)

15

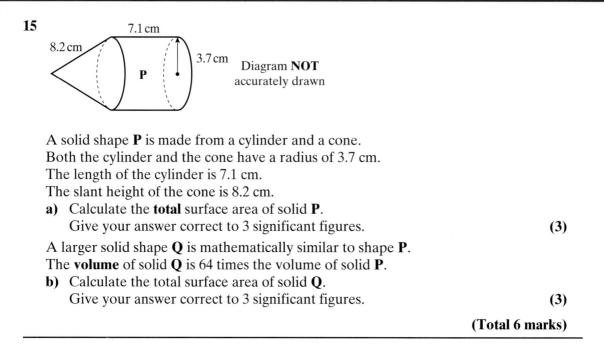

7.1 cm

8.2 cm

3.7 cm

P

Diagram **NOT**
accurately drawn

A solid shape **P** is made from a cylinder and a cone.
Both the cylinder and the cone have a radius of 3.7 cm.
The length of the cylinder is 7.1 cm.
The slant height of the cone is 8.2 cm.

a) Calculate the **total** surface area of solid **P**.
Give your answer correct to 3 significant figures. **(3)**

A larger solid shape **Q** is mathematically similar to shape **P**.
The **volume** of solid **Q** is 64 times the volume of solid **P**.

b) Calculate the total surface area of solid **Q**.
Give your answer correct to 3 significant figures. **(3)**

(Total 6 marks)

32 Advanced trigonometry

1

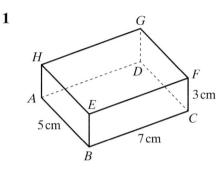

Diagram **NOT**
accurately drawn

ABCDEFGH is a cuboid of length 7 cm, width 5 cm and height 3 cm.
a) Calculate the size of the angle that *AF* makes with the plane *ABCD*.
Give your answer correct to 1 decimal place. **(4)**
b) Calculate the length of *AF*.
Give your answer correct to 3 significant figures. **(2)**

(Total 6 marks)

2

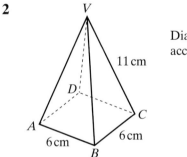

Diagram **NOT**
accurately drawn

The diagram shows a solid pyramid.
The base, *ABCD*, is a horizontal square of side 6 cm.
The vertex, *V*, is vertically above the midpoint of the base.
The length of each sloping edge is 11 cm.
Calculate the size of the angle the edge *VC* makes with the base.
Give your answer correct to 1 decimal place.

(Total 5 marks)

3

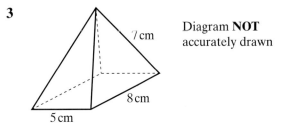

7 cm

8 cm

5 cm

Diagram **NOT**
accurately drawn

The diagram shows a pyramid.
The base is a horizontal rectangle of length 8 cm and width 5 cm.
The vertex is vertically above the midpoint of the base.
The length of each sloping edge is 7 cm.
Calculate the volume of the pyramid.
Give your answer correct to 3 significant figures.

(Total 6 marks)

4

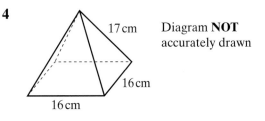

17 cm

16 cm

16 cm

Diagram **NOT**
accurately drawn

The diagram shows a solid pyramid.
The base is a horizontal square of side 16 cm.
The vertex is vertically above the midpoint of the base.
The length of each sloping edge is 17 cm.
Calculate the total surface area of the pyramid.

(Total 6 marks)

5

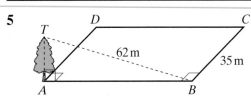

D

C

T

62 m

35 m

A

B

Diagram **NOT**
accurately drawn

ABCD is a horizontal rectangular field.
AT is a vertical tree.
BC = 35 m and *BT* = 62 m
The angle of elevation of *T* from *B* is 16°
Calculate the angle of elevation of *T* from *C*.
Give your answer correct to 1 decimal place.

(Total 6 marks)

6

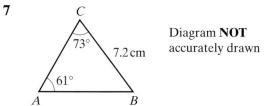

Diagram **NOT** accurately drawn

In triangle ABC, AB = 5.7 cm, AC = 6.8 cm and angle BAC = 58°
a) Calculate the area of the triangle.
   Give your answer correct to 3 significant figures. **(2)**
b) Calculate the length of BC.
   Give your answer correct to 3 significant figures. **(3)**

(Total 5 marks)

7

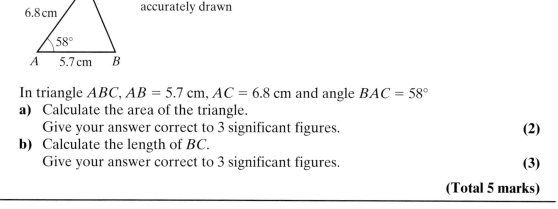

Diagram **NOT** accurately drawn

In triangle ABC, BC = 7.2 cm, angle BAC = 61° and angle ACB = 73°
Calculate the length of AB.
Give your answer correct to 3 significant figures.

(Total 3 marks)

8

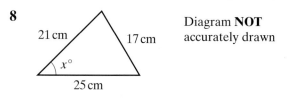

Diagram **NOT** accurately drawn

Calculate the value of x.
Give your answer correct to 1 decimal place.

(Total 3 marks)

9 The area of triangle ABC is 17 cm².
$AB - 6.9$ cm and $AC = 5.8$ cm
Calculate the size of the acute angle BAC.
Give your answer correct to 1 decimal place.

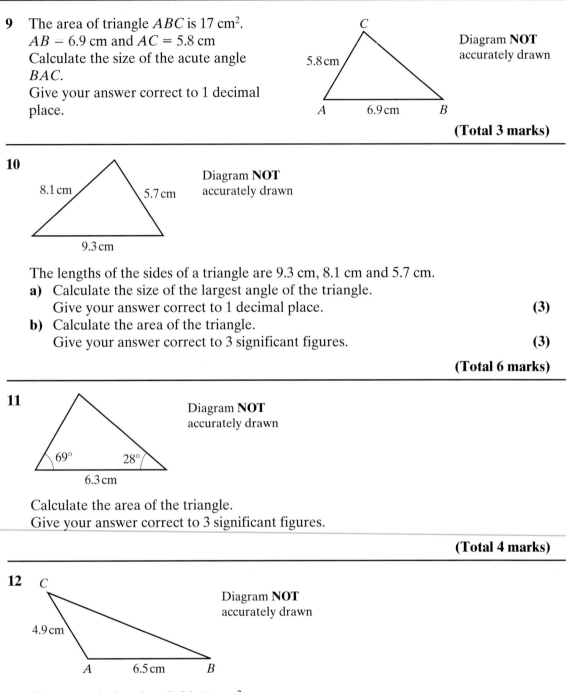

Diagram **NOT** accurately drawn

5.8 cm

6.9 cm

(Total 3 marks)

10

8.1 cm   5.7 cm

9.3 cm

Diagram **NOT** accurately drawn

The lengths of the sides of a triangle are 9.3 cm, 8.1 cm and 5.7 cm.
a) Calculate the size of the largest angle of the triangle.
   Give your answer correct to 1 decimal place. **(3)**
b) Calculate the area of the triangle.
   Give your answer correct to 3 significant figures. **(3)**

(Total 6 marks)

11

69°   28°

6.3 cm

Diagram **NOT** accurately drawn

Calculate the area of the triangle.
Give your answer correct to 3 significant figures.

(Total 4 marks)

12

C

4.9 cm

A   6.5 cm   *B*

Diagram **NOT** accurately drawn

The area of triangle ABC is 14 cm².
$AB = 6.5$ cm and $AC = 4.9$ cm
Calculate the size of the **obtuse** angle BAC.
Give your answer correct to 1 decimal place.

(Total 4 marks)

13

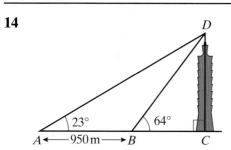

Diagram **NOT** accurately drawn

AC = 8.4 cm, angle BAC = 62° and angle ABC = 75°

a) Calculate the length of BC.
   Give your answer correct to 3 significant figures. **(3)**

b) Calculate the area of triangle ABC.
   Give your answer correct to 3 significant figures. **(3)**

(Total 6 marks)

14

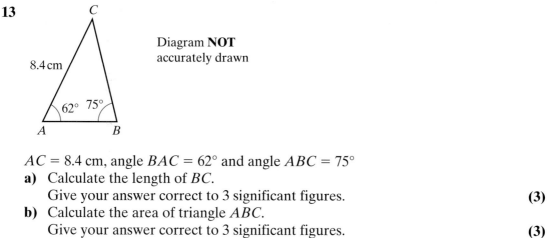

Diagram **NOT** accurately drawn

The diagram shows the Taipei 101 Tower in Taiwan.
The points A, B and C are in a straight line on horizontal ground.
The points C is at the foot of the tower and the point D is at the top of the tower.
AB = 950 m.
The angle of elevation of D from A is 23°
The angle of elevation of D from B is 64°
Calculate the height of the tower.
Give your answer correct to 3 significant figures.

(Total 5 marks)

15

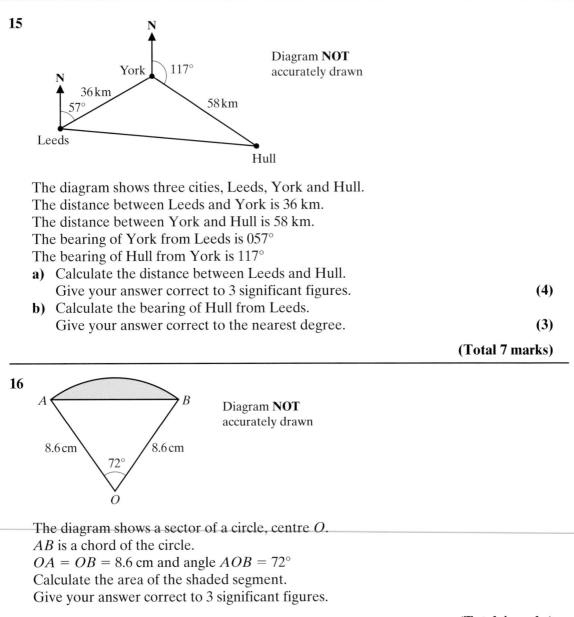

Diagram **NOT** accurately drawn

The diagram shows three cities, Leeds, York and Hull.
The distance between Leeds and York is 36 km.
The distance between York and Hull is 58 km.
The bearing of York from Leeds is 057°
The bearing of Hull from York is 117°
a) Calculate the distance between Leeds and Hull.
Give your answer correct to 3 significant figures. **(4)**
b) Calculate the bearing of Hull from Leeds.
Give your answer correct to the nearest degree. **(3)**

(Total 7 marks)

16

Diagram **NOT** accurately drawn

The diagram shows a sector of a circle, centre O.
AB is a chord of the circle.
$OA = OB = 8.6$ cm and angle $AOB = 72°$
Calculate the area of the shaded segment.
Give your answer correct to 3 significant figures.

(Total 4 marks)

33 Vectors

1 The point A has coordinates $(1, 2)$ and the point B has coordinates $(4, 7)$.

    a) i) Find $\overrightarrow{AB}$.

       ii) Find the magnitude of $\overrightarrow{AB}$.   **(3)**

    The point C has coordinates $(4, 1)$.

    D is a point such that $\overrightarrow{CD} = 2\,\overrightarrow{AB}$.

    b) Find the coordinates of the point D.   **(2)**

                                                    (Total 5 marks)

---

2 A is the point $(2, 3)$, B is the point $(4, 2)$, C is the point $(5, 7)$ and D is the point $(9, 5)$.

    a) i) Find $\overrightarrow{AB}$.

       ii) Find $\overrightarrow{CD}$.

      iii) Use your answers to parts **i)** and **ii)** to write down two geometric facts about AB and CD.   **(3)**

    b) Find the magnitude of $\overrightarrow{CD}$.   **(3)**

                                                    (Total 6 marks)

3

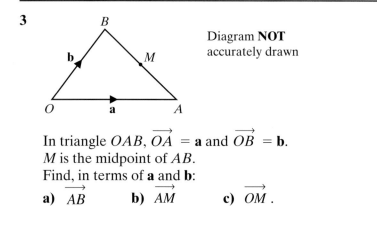

Diagram **NOT** accurately drawn

In triangle OAB, $\overrightarrow{OA} = \mathbf{a}$ and $\overrightarrow{OB} = \mathbf{b}$.

M is the midpoint of AB.

Find, in terms of $\mathbf{a}$ and $\mathbf{b}$:

    a) $\overrightarrow{AB}$ **b)** $\overrightarrow{AM}$ **c)** $\overrightarrow{OM}$.

                                                    (Total 3 marks)

4 PQR is a triangle.

X is a point on PQ such that $PX = \frac{1}{4}PQ$.

Y is a point on RQ such that $RY = \frac{1}{4}RQ$.

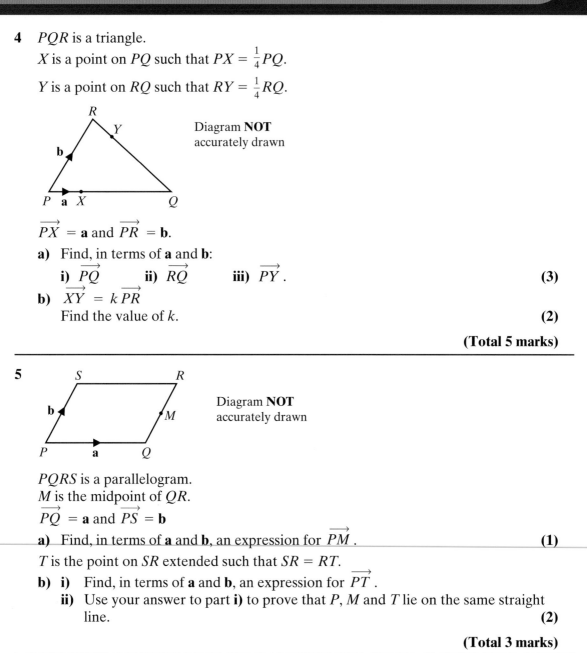

Diagram **NOT** accurately drawn

$\overrightarrow{PX} = \mathbf{a}$ and $\overrightarrow{PR} = \mathbf{b}$.

a) Find, in terms of $\mathbf{a}$ and $\mathbf{b}$:

    i) $\overrightarrow{PQ}$ **ii)** $\overrightarrow{RQ}$ **iii)** $\overrightarrow{PY}$.   **(3)**

b) $\overrightarrow{XY} = k\,\overrightarrow{PR}$

    Find the value of k.   **(2)**

(Total 5 marks)

5

Diagram **NOT** accurately drawn

$PQRS$ is a parallelogram.

M is the midpoint of QR.

$\overrightarrow{PQ} = \mathbf{a}$ and $\overrightarrow{PS} = \mathbf{b}$

a) Find, in terms of $\mathbf{a}$ and $\mathbf{b}$, an expression for $\overrightarrow{PM}$.   **(1)**

T is the point on SR extended such that $SR = RT$.

b) i) Find, in terms of $\mathbf{a}$ and $\mathbf{b}$, an expression for $\overrightarrow{PT}$.

    ii) Use your answer to part **i)** to prove that P, M and T lie on the same straight line.   **(2)**

(Total 3 marks)

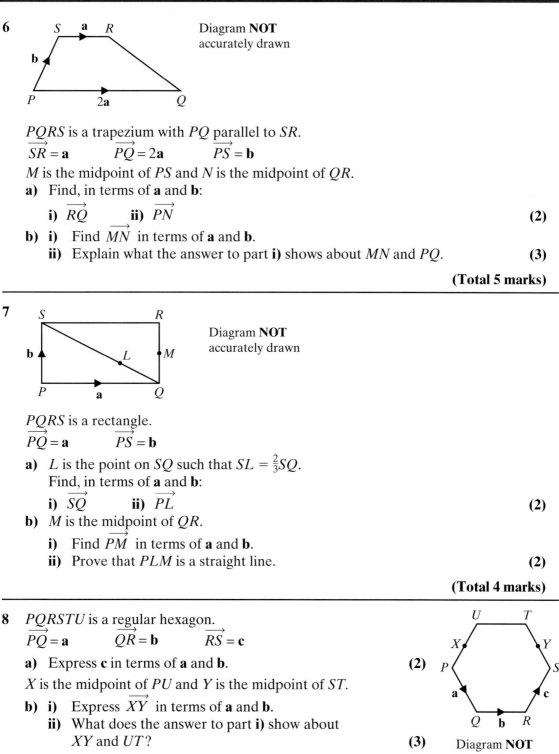

6

PQRS is a trapezium with *PQ* parallel to *SR*.
$\overrightarrow{SR} = \mathbf{a}$ $\overrightarrow{PQ} = 2\mathbf{a}$ $\overrightarrow{PS} = \mathbf{b}$
M is the midpoint of *PS* and *N* is the midpoint of *QR*.
a) Find, in terms of **a** and **b**:
  i) $\overrightarrow{RQ}$ **ii)** $\overrightarrow{PN}$ **(2)**
b) i) Find $\overrightarrow{MN}$ in terms of **a** and **b**.
  ii) Explain what the answer to part **i)** shows about *MN* and *PQ*.   **(3)**

(Total 5 marks)

7

PQRS is a rectangle.
$\overrightarrow{PQ} = \mathbf{a}$ $\overrightarrow{PS} = \mathbf{b}$
a) *L* is the point on *SQ* such that $SL = \frac{2}{3}SQ$.
  Find, in terms of **a** and **b**:
  i) $\overrightarrow{SQ}$ **ii)** $\overrightarrow{PL}$ **(2)**
b) *M* is the midpoint of *QR*.
  i) Find $\overrightarrow{PM}$ in terms of **a** and **b**.
  ii) Prove that *PLM* is a straight line.   **(2)**

(Total 4 marks)

8 *PQRSTU* is a regular hexagon.
$\overrightarrow{PQ} = \mathbf{a}$ $\overrightarrow{QR} = \mathbf{b}$ $\overrightarrow{RS} = \mathbf{c}$
a) Express **c** in terms of **a** and **b**.   **(2)**
X is the midpoint of *PU* and *Y* is the midpoint of *ST*.
b) i) Express $\overrightarrow{XY}$ in terms of **a** and **b**.
  ii) What does the answer to part **i)** show about *XY* and *UT*?   **(3)**

(Total 5 marks)

145

9

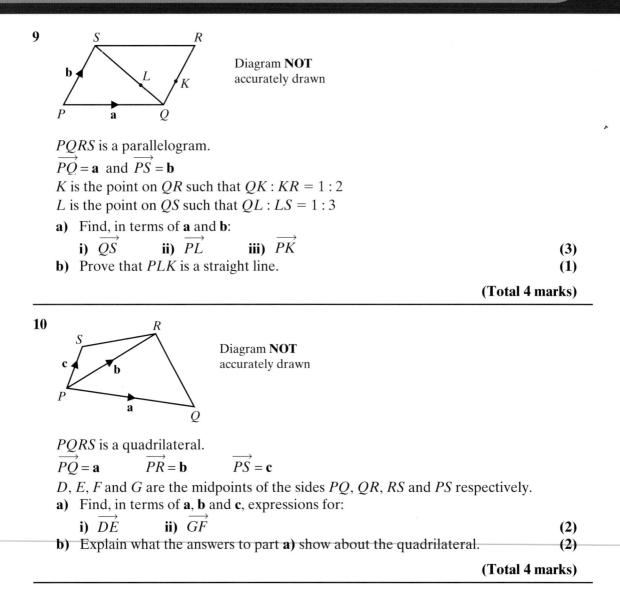

Diagram **NOT** accurately drawn

$PQRS$ is a parallelogram.

$\overrightarrow{PQ} = \mathbf{a}$ and $\overrightarrow{PS} = \mathbf{b}$

K is the point on QR such that $QK : KR = 1 : 2$

L is the point on QS such that $QL : LS = 1 : 3$

a) Find, in terms of $\mathbf{a}$ and $\mathbf{b}$:

    i) $\overrightarrow{QS}$ **ii)** $\overrightarrow{PL}$ **iii)** $\overrightarrow{PK}$ **(3)**

b) Prove that PLK is a straight line.   **(1)**

(Total 4 marks)

10

Diagram **NOT** accurately drawn

$PQRS$ is a quadrilateral.

$\overrightarrow{PQ} = \mathbf{a}$ $\overrightarrow{PR} = \mathbf{b}$ $\overrightarrow{PS} = \mathbf{c}$

D, E, F and G are the midpoints of the sides PQ, QR, RS and PS respectively.

a) Find, in terms of $\mathbf{a}, \mathbf{b}$ and $\mathbf{c}$, expressions for:

    i) $\overrightarrow{DE}$ **ii)** $\overrightarrow{GF}$ **(2)**

b) Explain what the answers to part **a)** show about the quadrilateral.   **(2)**

(Total 4 marks)

Handling Data

34 Graphical representation of data

1   The cumulative frequency table gives information about the heights of 80 men.

Height (h cm)	Cumulative frequency
$140 < h \leqslant 150$	4
$140 < h \leqslant 160$	14
$140 < h \leqslant 170$	34
$140 < h \leqslant 180$	68
$140 < h \leqslant 190$	80

a)   Work out the number of men with heights in the interval $150 < h \leqslant 160$ **(1)**

b)   Work out the number of men with heights greater than 170 cm.   **(1)**

c)   On a copy of the grid, draw the cumulative frequency graph for the table.

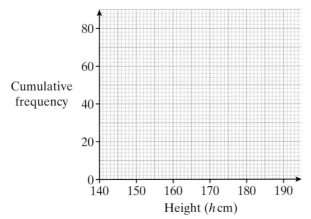

   (2)

d)   Use your graph to find an estimate for the number of men with a height of less than 176 cm.   **(2)**

   (Total 6 marks)

2 The grouped frequency table gives information about the ages of the 195 members of a sports club.

Age (*t* years)	Frequency
$10 < t \leqslant 20$	20
$20 < t \leqslant 30$	95
$30 < t \leqslant 40$	65
$40 < t \leqslant 50$	10
$50 < t \leqslant 60$	5

a) Copy and complete the cumulative frequency table.

Age (*t* years)	Cumulative frequency
$10 < t \leqslant 20$	
$10 < t \leqslant 30$	
$10 < t \leqslant 40$	
$10 < t \leqslant 50$	
$10 < t \leqslant 60$	

(1)

b) On a copy of the grid, draw the cumulative frequency graph for your table.

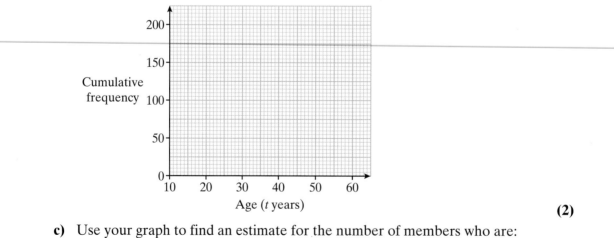

(2)

c) Use your graph to find an estimate for the number of members who are:
   i) aged less than 26
   ii) aged more than 35

(3)

(Total 6 marks)

3   The cumulative frequency graph gives information about the weights, in kilograms, of the sheep in a flock.

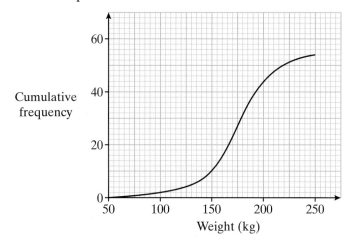

Weight (kg)

Use the cumulative frequency graph to find:
a)   the number of sheep in the flock
b)   an estimate for the number of sheep that weighed less than 145 kg
c)   an estimate for the number of sheep that weighed more than 210 kg.

(Total 4 marks)

4   The unfinished table and histogram give information about the times taken by some students to complete a homework task.

Time (t minutes)	Frequency
$0 < t \leqslant 15$	30
$15 < t \leqslant 20$	
$20 < t \leqslant 30$	
$30 < t \leqslant 55$	55
$55 < t \leqslant 65$	16

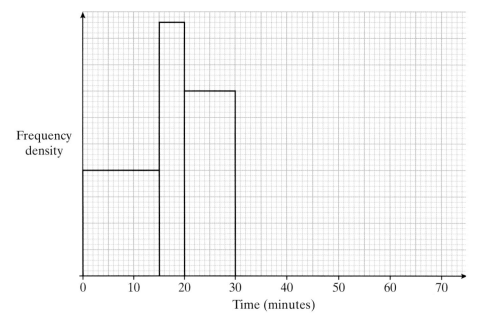

a) Use the information in the table to complete a copy of the histogram. **(2)**
b) Use the information in the histogram to complete a copy of the table. **(2)**

(Total 4 marks)

5 The histogram gives information about the lengths of some pieces of seaweed.

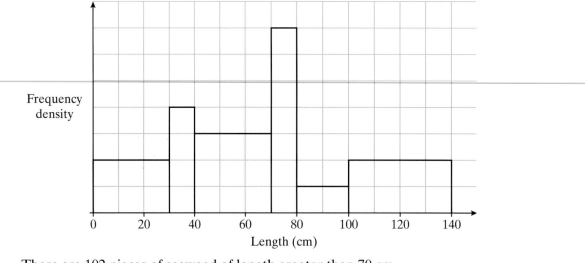

There are 102 pieces of seaweed of length greater than 70 cm.
a) How many pieces of seaweed have a length greater than 1 metre? **(2)**
b) Calculate the total number of pieces of seaweed. **(2)**

(Total 4 marks)

6 The histogram shows information about the masses of some apples.

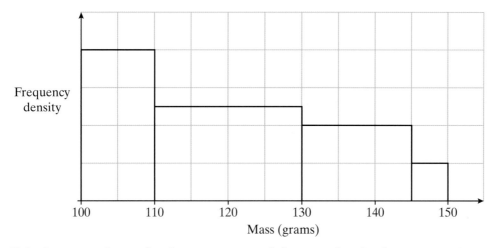

Mass (grams)

Calculate an estimate for the percentage of these apples that have a mass greater than 140 grams.

(Total 3 marks)

7 The histogram shows information about the heights, h cm, of some Year 11 students.

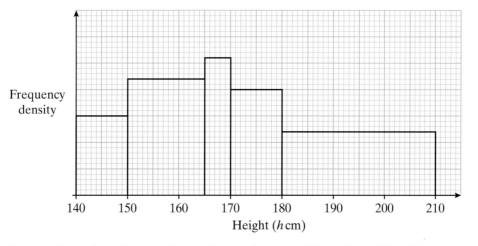

Height (h cm)

The number of students with heights in the class $170 \leqslant h < 180$ is 20
a) Find the number of students with heights in the class:
    i) $140 \leqslant h < 150$
    ii) $150 \leqslant h < 165$ **(3)**
b) Calculate an estimate of the number of these Year 11 students with heights from 175 cm to 190 cm.   **(3)**

(Total 6 marks)

35 Statistical measures

1 The table shows information about the ages, in years, of 20 children at a party:

Age in years	Frequency
3	2
4	12
5	6

    a) Work out the range of their ages.   **(1)**
    b) Work out the mean age of the children.   **(3)**

(Total 4 marks)

2 The table shows information about the number of empty seats on 40 coaches arriving at a football stadium:

Number of empty seats	Frequency
0	20
1	3
2	8
3	5
4	4

    a) Write down the mode of the number of empty seats.   **(1)**
    b) Work out the mean number of empty seats per coach.   **(3)**

(Total 4 marks)

3 25 students scored goals for the Year 11 football team last season.
The table shows information about the number of goals they scored:

Number of goals scored	Number of students
1	10
2	4
3	6
4	1
5	4

Work out the total number of goals the team scored last season.

(Total 3 marks)

4 Zandile buys 20 boxes of drawing pins. She counts the number of drawing pins in each box. The table shows information about her results:

Number of drawing pins	Frequency
28	4
29	5
30	3
31	8

a) i) Find the median number of drawing pins.
ii) Calculate the mean number of drawing pins in the 20 boxes. **(5)**
Zandile is given another box containing 30 drawing pins.
b) i) Without calculating the new mean, state whether the mean will increase or decrease or stay the same.
ii) Give a reason for your answer to **b i)**. **(2)**

(Total 7 marks)

5 a) Five numbers have a mean of 8
Four of the numbers are 2, 7, 9 and 10
Find the other number. **(2)**
b) Three numbers have a mode of 8 and a mean of 6
Find the three numbers. **(2)**
c) Find four numbers which have a mode of 4 and a median of 3 **(2)**

(Total 6 marks)

6 A class of 18 boys and 12 girls sit a Science exam.
The mean mark for the boys is 62.5%
The mean mark for the girls is 58.75%
Work out the mean mark for the class.

(Total 3 marks)

7 The mean age of 5 people in a car is 24 years.
The age of the driver is 36 years.
Work out the mean age of the 4 passengers.

(Total 3 marks)

8 The mean height of a group of six boys is 162 cm.
Four of the boys have a mean height of 159 cm.
Work out the mean height of the other two boys in the group.

(Total 4 marks)

9 Adil writes down ten numbers.
The mean of the ten numbers is 17.2
What extra number must Adil write down so that the mean of the eleven numbers is 18?

(Total 3 marks)

10 The grouped frequency table gives information about the daily rainfall in Mathstown last June:

Daily rainfall (d mm)	Number of days
$0 \leqslant d < 2$	12
$2 \leqslant d < 4$	5
$4 \leqslant d < 6$	6
$6 \leqslant d < 8$	3
$8 \leqslant d < 10$	4

a) Write down the modal class. **(1)**
b) In which interval does the median lie? **(1)**
c) Calculate an estimate for the total rainfall last June. **(3)**

(Total 5 marks)

11 The table gives information about the speeds, in km/h, of 50 cars passing a speed checkpoint on a motorway:

Speed (v km/h)	Frequency
$70 < v \leq 80$	4
$80 < v \leq 90$	9
$90 < v \leq 100$	15
$100 < v \leq 110$	16
$110 < v \leq 120$	6

a) Write down the modal class. **(1)**
b) Work out an estimate for the probability that the next car passing the speed checkpoint will have a speed of more than 100 km/h. **(2)**
c) Work out an estimate for the mean speed of the 50 cars. **(4)**

(Total 7 marks)

12 The cumulative frequency graph gives information about the time, in hours, spent on the internet last week by each of 80 students:

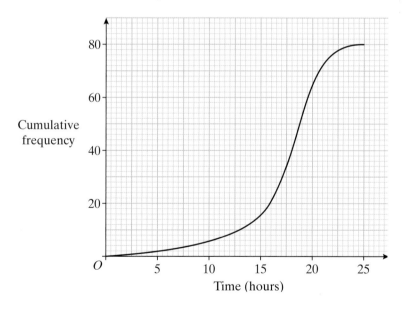

a) Find an estimate for the median of the times. **(2)**
b) Find an estimate for the interquartile range of the times. **(2)**

(Total 4 marks)

13 The table gives information about the number of copies of *The Sunday Headline* sold each Sunday last year:

Number (n) of copies sold	Frequency
$20\,000 < n \leqslant 30\,000$	4
$30\,000 < n \leqslant 40\,000$	16
$40\,000 < n \leqslant 50\,000$	16
$50\,000 < n \leqslant 60\,000$	10
$60\,000 < n \leqslant 70\,000$	4
$70\,000 < n \leqslant 80\,000$	2

a) Work out an estimate for the mean number of copies sold each Sunday. **(4)**

b) Copy and complete the cumulative frequency table.

Number (n) of copies sold	Cumulative frequency
$20\,000 < n \leqslant 30\,000$	
$20\,000 < n \leqslant 40\,000$	
$20\,000 < n \leqslant 50\,000$	
$20\,000 < n \leqslant 60\,000$	
$20\,000 < n \leqslant 70\,000$	
$20\,000 < n \leqslant 80\,000$	

(1)

c) On a copy of the grid, draw the cumulative frequency graph for your table.

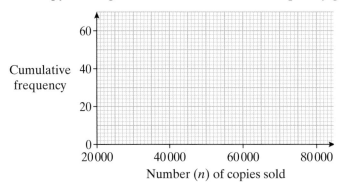

(2)

d) Use your graph to find an estimate for:
   i) the median number of copies sold
   ii) the interquartile range of the number of copies sold. **(3)**

(Total 10 marks)

14 Here are the ages, in years, of the 11 players in a football team:

    21 26 18 29 20 26 25 19 28 23 21

  a) Find
      i) the lower quartile of these ages
      ii) the upper quartile of these ages.   **(3)**

  The ages of the 11 players in a cricket team have an interquartile range of 19 years.
  b) Make one comparison between the ages of the two teams.   **(1)**

                                                              (Total 4 marks)

15 All the students in a class took a French test.
  There are 15 boys in the class.
  The boys gained these marks in the test:

    11 5 14 20 18 4 8 11 18 12 13 15 10 17 7

  a) Find the interquartile range of the boys' marks.   **(3)**

  In the test, the marks of the girls in the class had a median of 14 and an interquartile range of 6
  b) Make two comparisons between the boys' marks and the girls' marks.   **(2)**

                                                              (Total 5 marks)

16 180 students took an English examination.
  The table shows information about their marks:

Mark (x)	Frequency
$0 \leqslant x < 20$	15
$20 \leqslant x < 40$	95
$40 \leqslant x < 60$	45
$60 \leqslant x < 80$	15
$80 \leqslant x < 100$	10

  a) Copy and complete the cumulative frequency table.

Mark (x)	Cumulative frequency
$0 \leqslant x < 20$	
$0 \leqslant x < 40$	
$0 \leqslant x < 60$	
$0 \leqslant x < 80$	
$0 \leqslant x < 100$	

                                                                     (1)

b) On a copy of the grid, draw the cumulative frequency graph for your table.

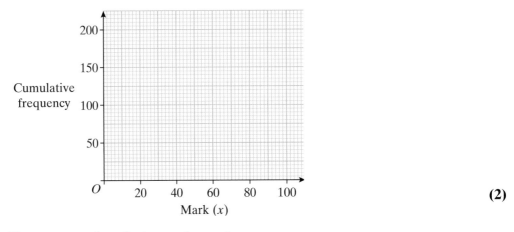

(2)

c) Use your graph to find an estimate for:
    i) the median mark
    ii) the interquartile range of the marks. **(3)**
d) 145 students passed the examination.
    Use your graph to find an estimate of the pass mark for the examination. **(2)**

(Total 8 marks)

17 The time each of 80 people took to travel to work was recorded.
The lower quartile of their times was 18 minutes.
Their median time was 28 minutes.
The upper quartile of their times was 35 minutes.
Find an estimate for the number of people who
a) took less than 25 minutes **(3)**
b) took more than 30 minutes. **(2)**
You may use a copy of the grid to find your estimates.

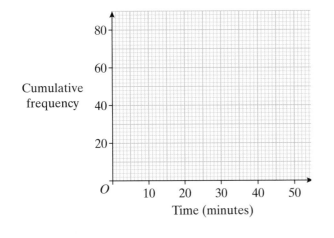

(Total 5 marks)

18 The frequency diagram gives information about the marks gained by a class of 30 students in an examination:

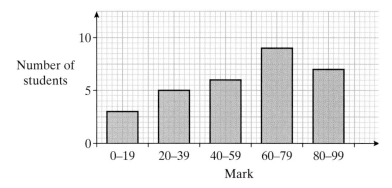

a) Write down the modal class. **(1)**
b) Work out an estimate for the mean mark of the 30 students. **(4)**
c) On a copy of the grid, draw a cumulative frequency graph for the marks.

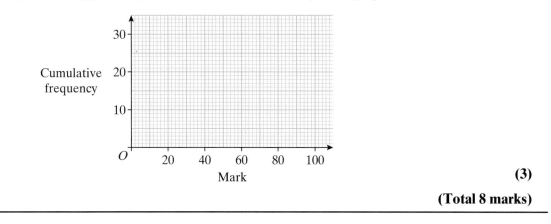

(3)

(Total 8 marks)

36 Probability

1   Cara spins a coin 100 times. It comes down heads 75 times.
   a) Find an estimate of the probability that it will come down heads the next time
        she spins it.   **(1)**
   b) Is the coin fair? Give a reason for your answer.   **(1)**

                                                                            (Total 2 marks)

2   The probability that a person chosen at random is left-handed is 0.1
   350 people are to be chosen at random.
   Work out an estimate for the number of people who will be left-handed.

                                                                            (Total 2 marks)

3   A bag contains red beads, white beads and blue beads.
   Kevin takes a bead at random and replaces it.
   He does this 80 times.
   The table shows the number of times he took a bead of each colour:

Colour	Red	White	Blue
Number of times	19	48	13

   There are 35 beads in the bag.
   Work out an estimate for the number of white beads in the bag.

                                                                            (Total 3 marks)

4   Here is a three-sided spinner.
   Its sides are labelled 1, 2 and 3
   The spinner is biased.
   The probability that the spinner lands on the number 1 is $\frac{3}{8}$
   Mary is going to spin the spinner 48 times.
   Work out an estimate for the number of times the spinner lands on 1

                                                                            (Total 2 marks)

5 Here are ten polygons:

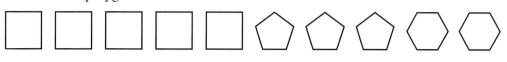

Five of the polygons are squares.
Three of the polygons are pentagons.
Two of the polygons are hexagons.
The polygons are put in a bag.
David takes a polygon at random from the bag 120 times.
He replaces the polygon each time.
Work out an estimate for the number of times he will take:
a) a pentagon (2)
b) a polygon with less than 6 sides. (2)

(Total 4 marks)

6 Four athletes Abi, Brendan, Christopher and Dale take part in a race.
The table shows the probabilities that Abi or Brendan or Christopher will win the race:

Abi	Brendan	Christopher	Dale
0.1	0.25	0.3	

a) Work out the probability that Brendan or Christopher will win the race. (2)
b) Work out the probability that Dale will win the race. (2)

(Total 4 marks)

7 Here is a four-sided spinner.
The sides of the spinner are labelled 2, 3, 4 and 5
The spinner is biased.
The probability that the spinner will land on each of the
numbers 2, 3 and 4 is given in the table:

Number	2	3	4	5
Probability	0.35	0.15	0.3	

The spinner is spun once.
a) Work out the probability that the spinner lands on 5 (2)
b) Work out the probability that the spinner lands on an even number. (2)

Oscar is going to spin the spinner 80 times.
c) Work out an estimate for the number of times the spinner will land on 3 (2)

(Total 6 marks)

8 A box contains some shapes.
The colour of each shape is black or green or red or white.
A shape is taken at random from the box.
The table shows the probability that the shape is black or green or red:

Colour	Probability
Black	0.16
Green	0.4
Red	0.2
White	

a) Work out the probability that the shape is white. **(2)**

b) Work out the probability that the shape is black or green. **(2)**

The probability that the shape is a square is 0.7

c) Brody says:
'*The probability that the shape is a black square is 0.16 + 0.7 = 0.86*'
Is Brody correct?
Give a reason for your answer. **(2)**

(Total 6 marks)

9 Ami, Bea, Carla and Deano play a game.
There can only be one winner.
The probability that Ami will win is 0.25
The probability that Bea will win is 0.15

a) Calculate the probability that either Carla or Deano will win. **(2)**

Carla is three times more likely to win than Deano.

b) Calculate the probability that Carla will win. **(2)**

(Total 4 marks)

10 Stuart is going to play one game of tennis and one game of badminton.
   The probability that he will win the game of tennis is $\frac{2}{3}$. The probability that he will win the game of badminton is $\frac{3}{5}$

   a) Complete a copy of the probability tree diagram to show this information.

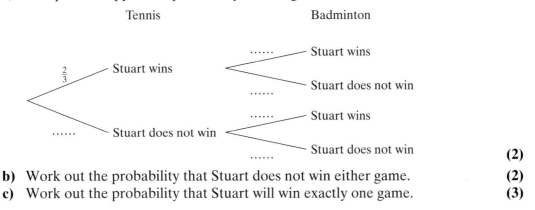

   b) Work out the probability that Stuart does not win either game.   (2)
   c) Work out the probability that Stuart will win exactly one game.   (3)

   (Total 7 marks)

11 A bag contains 5 black discs and 3 white discs.
   Saira takes a disc at random from the bag and notes its colour.
   She then replaces the disc in the bag.
   Saira takes a second disc at random from the bag and notes its colour.
   a) Complete a copy of the probability tree diagram.

   First disc   Second disc

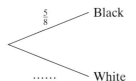
                                                                    (2)
   b) Calculate the probability that Saira takes two white discs.   (2)
   c) Calculate the probability that Saira takes two discs of different colours.   (3)

   (Total 7 marks)

12 In a television quiz show, a contestant must answer one question correctly in the first round to go through to the second round.
   In the first round each contestant is asked, at most, two questions.
   A contestant who answers the first question correctly goes through to the second round and is not asked a second question.
   A contestant who does not answer the first question correctly is asked a second question. If they answer the second question correctly, they go through to the second round. If they do not answer the second question correctly, they are eliminated.
   The probability that a contestant will answer a question correctly is 0.6

a) Complete a copy of the probability tree diagram by writing the appropriate probability on each of the four branches.

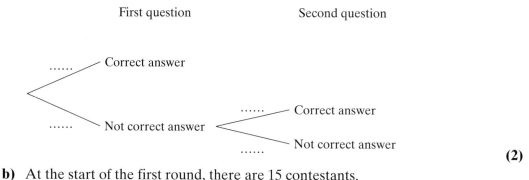

First question   Second question

(2)

b) At the start of the first round, there are 15 contestants.
Work out an estimate for the number of contestants who will go through to the second round.

(5)

(Total 7 marks)

---

13 Nick has 10 ties in a bag.
6 of the ties are red and 4 of the ties are black.
Nick takes two ties from the bag at random, **without** replacement.
a) Complete a copy of the probability tree diagram.

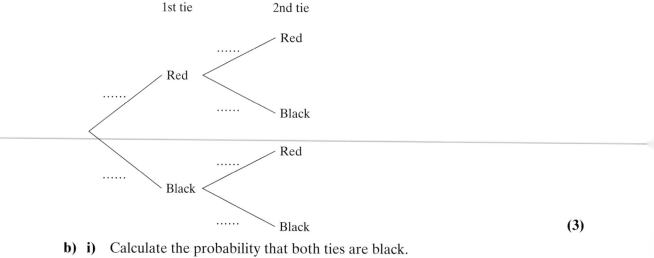

(3)

b) i) Calculate the probability that both ties are black.
   ii) Calculate the probability that the ties are different colours.

(4)

(Total 7 marks)

14

John has six coins in his pocket.
He has one 50 cent coin, three 20 cent coins and two 10 cent coins.
He takes a coin at random from his pocket and records its value.
He then puts the coin back in his pocket.
John takes a second coin at random from his pocket and records its value.
a) Complete a copy of the probability tree diagram.

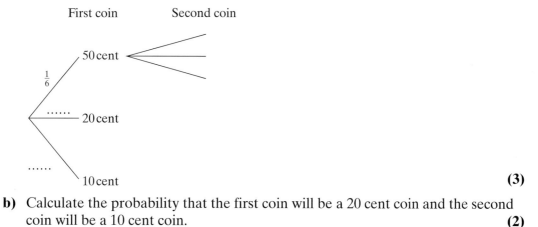

(3)

b) Calculate the probability that the first coin will be a 20 cent coin and the second coin will be a 10 cent coin. **(2)**
c) Calculate the probability that the two coins will have the same value. **(4)**

(Total 9 marks)

15 In order to become a fireman, Adil has to pass a test.
He is allowed only three attempts to pass the test.
The probability that Adil will pass the test at his first attempt is $\frac{3}{8}$
If he fails at his first attempt, the probability that he will pass at his second attempt is $\frac{3}{5}$
If he fails at his second attempt, the probability that he will pass at his third attempt is $\frac{2}{3}$
a) Complete a copy of the probability tree diagram.

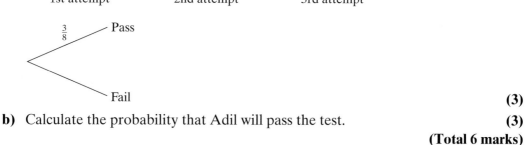

(3)

b) Calculate the probability that Adil will pass the test. **(3)**

(Total 6 marks)

16 Meg sits a geography test and a biology test.

The probability that she will pass the geography test is $\frac{2}{3}$

The probability that she will pass the biology test is $\frac{5}{8}$

a) Calculate the probability that she will not pass either test. **(2)**

b) Calculate the probability that she will pass exactly one test. **(3)**

(Total 5 marks)

---

17 The diagram shows eight counters:

$$\text{P} \quad \text{A} \quad \text{R} \quad \text{A} \quad \text{L} \quad \text{L} \quad \text{E} \quad \text{L}$$

Each counter has a letter on it.

Will puts the eight counters into a bag.

He takes a counter at random from the bag and does not replace it.

He then takes a second counter at random from the bag.

a) Calculate the probability that both letters he takes will be L. **(2)**

b) Calculate the probability that at least one of the letters he takes will be A. **(3)**

(Total 5 marks)

---

18 Here is a biased spinner:

When the spinner is spun, the score is 1 or 3 or 5

The probability that the score is 1 is 0.4

The probability that the score is 3 is 0.5

Kezia spins the spinner twice.

a) Work out the probability that the sum of the two scores is 2 **(2)**

b) Work out the probability that the sum of the two scores is 6 **(3)**

(Total 5 marks)

---

19 Farah has two bags.

The first bag contains 4 red beads and 1 white bead.

The second bag contains 5 red beads and 3 white beads.

Farah takes one bead at random from each bag.

a) Calculate the probability that Farah takes two white beads. **(2)**

b) Calculate the probability that Farah takes one bead of each colour. **(3)**

(Total 5 marks)

---

20 A tin of biscuits contains 15 chocolate biscuits and 10 cream biscuits.
Silvia takes a biscuit at random from the tin and eats it.
She then takes another biscuit at random from the tin and eats it.
  a) Calculate the probability that both the biscuits Silvia eats are chocolate. **(2)**
  b) Calculate the probability that at least one of the biscuits Silvia eats is cream. **(3)**

(Total 5 marks)

21 Harry has a biased coin.
When he spins the coin, the probability that it will come down tails is 0.7
Harry spins the coin three times.
  a) Calculate the probability that it comes down **heads** every time. **(2)**
  b) Calculate the probability that it comes down tails at least twice. **(4)**

(Total 6 marks)

22 Mandy has 20 stamps in an envelope.
10 stamps are German.
8 stamps are Spanish.
2 stamps are English.
Mandy takes three stamps from the envelope at random, **without** replacement.
  a) Calculate the probability that Mandy takes both English stamps. **(3)**
  b) Calculate the probability that she takes stamps from at least two countries. **(3)**

(Total 6 marks)

23 A fair six-sided dice has faces numbered 1, 2, 3, 4, 5 and 6
When a dice is thrown the number facing up is the score.
The dice is thrown three times.
Calculate the probability that the total score is 5

(Total 4 marks)

24 A weekend tennis tournament is due to take place next Saturday and Sunday.
If rain interrupts play on the Saturday, the probability that it will interrupt play on the Sunday is 0.75
If rain does not interrupt play on the Saturday, the probability that it will interrupt play on the Sunday is 0.2
The probability that rain will interrupt play on the Saturday is 0.6
  a) Calculate the probability that rain will **not** interrupt play during the tournament. **(3)**
  b) Calculate the probability that rain will interrupt play on the Sunday. **(3)**

(Total 6 marks)

25 There are n marbles in a bag.

Five of the marbles are green and the rest are yellow.

Alison takes one marble at random from the bag.

She records its colour and then replaces it.

She takes a second marble at random from the bag and records its colour.

The probability that the two marbles Alison takes will have different colours is $\frac{3}{8}$

Show that $3n^2 - 80n + 400 = 0$

(Total 4 marks)

---

26 30 students were asked if they had eaten an Apple (A) or a Banana (B) or both in the past week.

The Venn diagram shows some of the results.

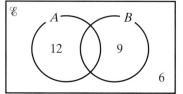

One of the 30 students is selected at random.

a) Find the probability that the student had eaten an apple but not a banana. **(1)**

b) Find the probability that the student had eaten an apple and a banana. **(1)**

c) Given that the student had eaten only one of the two fruits, find the probability that the student had eaten a banana. **(1)**

(Total 3 marks)

---

37 Further problem solving and mathematical reasoning

1   Eagles, Falcons, Hawks and Kestrels are four soccer teams.
    The four teams play in a tournament in which each team plays one match against every other team.
    A team receives 2 points for each match won, 1 point for each match drawn and 0 points for each match lost.
    The results of the matches in the tournament are used to complete this table.

	Played	Won	Drawn	Lost	Goals		Points
					For	Against	
Hawks	3	2	1	0	2	0	5
Falcons	3	2	0	1	5	4	4
Kestrels	3	1	1	1	4	4	3
Eagles	3	0	0	3	0	3	0

Find the score in the match between Falcons and Kestrels.

(Total 4 marks)

2   A bag contains beads.
    Each bead is red, blue or green.
    Esther takes at random one bead from the bag.
    The probability that she will take a red bead is 0.35
    The probability that she will take a blue bead is $\frac{3}{8}$

    Find the smallest possible number of beads which could be in the bag.

(Total 3 marks)

3   Amy, Bob and Chris are three swimmers.
    Amy swims a length of a swimming pool in 15 seconds.
    Bob swims a length of the swimming pool in 20 seconds.
    Chris swims a length of the swimming pool in 25 seconds.
    They start at the same time from the same end of the swimming pool.
    They continue swimming until they are all at the same end again.
    After how many seconds does this happen?

(Total 4 marks)

4   A crate is a cube with side 1 m.
    A box is a cuboid which is 40 cm by 20 cm by 10 cm.
    Find the greatest number of boxes which can be packed in the crate.

(Total 4 marks)

5   *ABCD* is a trapezium.
    AB is parallel to *DC*.
    AD = 9.6 cm, *BC* = 10.4 cm and *CD* = 5.7 cm.
    Angle *BAD* = 72°
    Calculate the length of *AB*.
    Give your answer correct to 3 significant figures.

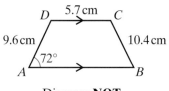

Diagram **NOT**
accurately drawn

(Total 6 marks)

6

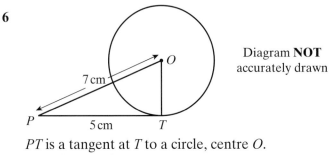

Diagram **NOT**
accurately drawn

PT is a tangent at *T* to a circle, centre *O*.
PT = 5 cm and *OP* = 7 cm.
Calculate the area of the circle.
Give your answer correct to 3 significant figures.

(Total 4 marks)

7

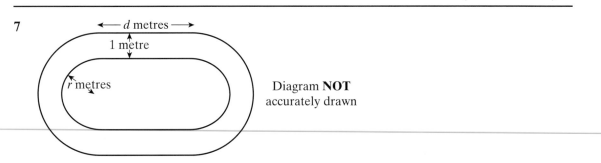

Diagram **NOT**
accurately drawn

The diagram shows the markings of a running track lane.
The inside of the lane consists of two straight lines, each of length *d* metres, and two semicircles, each of radius *r* metres.
The perimeter of the inside of the lane is 400 metres.
a) i) Explain why $2\pi r + 2d = 400$
The lane is 1 metre wide.
The outside of the lane consists of two straight lines, each of length *d* metres, and two semicircles.
    ii) Calculate the perimeter of the outside of the lane.
             Give your answer correct to 3 significant figures.   **(5)**

When the width of the lane is w metres, the perimeter of the outside of the lane is P metres.

b) Find, in terms of w, a formula for P.
Give your answer as simply as possible. **(2)**

(Total 7 marks)

8

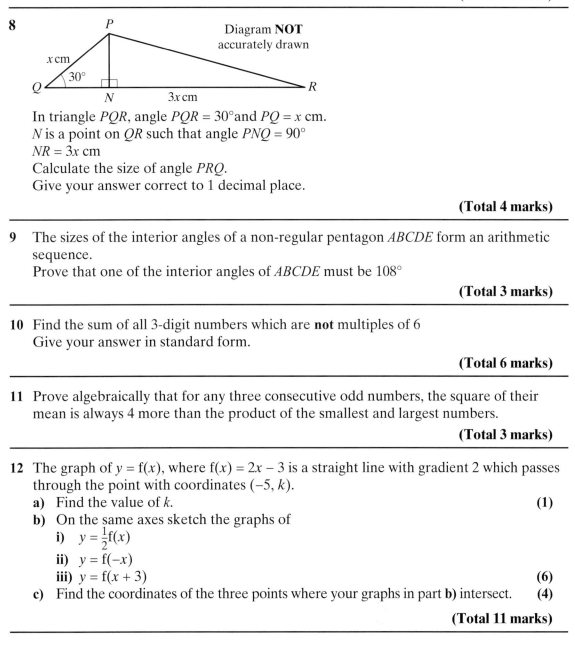

In triangle PQR, angle $PQR = 30°$ and $PQ = x$ cm.
N is a point on QR such that angle $PNQ = 90°$
$NR = 3x$ cm
Calculate the size of angle PRQ.
Give your answer correct to 1 decimal place.

(Total 4 marks)

---

9 The sizes of the interior angles of a non-regular pentagon $ABCDE$ form an arithmetic sequence.
Prove that one of the interior angles of $ABCDE$ must be $108°$

(Total 3 marks)

---

10 Find the sum of all 3-digit numbers which are **not** multiples of 6
Give your answer in standard form.

(Total 6 marks)

---

11 Prove algebraically that for any three consecutive odd numbers, the square of their mean is always 4 more than the product of the smallest and largest numbers.

(Total 3 marks)

---

12 The graph of $y = f(x)$, where $f(x) = 2x - 3$ is a straight line with gradient 2 which passes through the point with coordinates $(-5, k)$.
a) Find the value of k. **(1)**
b) On the same axes sketch the graphs of
   i) $y = \frac{1}{2}f(x)$
   ii) $y = f(-x)$
   iii) $y = f(x + 3)$ **(6)**
c) Find the coordinates of the three points where your graphs in part **b)** intersect. **(4)**

(Total 11 marks)

13 The curve with equation $y = 7 - 2x - x^2$ has a single stationary point whose coordinates are $(-1, 8)$.
A vertical translation is applied to the curve to give the curve $y = g(x)$
The stationary point of the curve $y = g(x)$ lies on the x-axis.
a) Find a quadratic expression for $g(x)$. **(2)**
A horizontal translation is applied to the curve $y = 7 - 2x - x^2$ to give the curve $y = h(x)$.
The stationary point of the curve $y = h(x)$ lies on the y-axis.
b) Find a quadratic expression for $h(x)$. **(2)**

(Total 4 marks)

14 a) i) Sketch the graph of the curve with equation $y = \sin x°$ for $-180 \leqslant x \leqslant 180$
ii) Write down the coordinates of the turning points of your graph. **(4)**
b) i) Sketch the graph of the curve with equation $y = \cos x°$ for $-240 \leqslant x \leqslant 240$
ii) Write down the coordinates of the turning points of your graph. **(4)**
c) The graph of $y = \sin x°$ can be mapped onto the graph of $y = \cos x°$ by a translation $\begin{pmatrix} c \\ 0 \end{pmatrix}$. Write down a possible value for c. **(1)**

(Total 9 marks)

15 Here are three statements:
I The equation $\sin x° + \cos x° = 3$ has a solution in the interval $-180 \leqslant x \leqslant 180$
II The equation $\tan x° + \cos x° = 2$ has a solution in the interval $-90 < x \leqslant 0$
III The equation $\tan x° - \sin x° = 0.6$ has a solution in the interval $90 < x < 180$
For each statement, explain why the statement cannot be true.

(Total 6 marks)

16 A curve has equation $y = 4x^2 - 16x + 19$
The curve crosses the y-axis at the point P.
The curve has a turning point T.
a) Write $4x^2 - 16x + 19$ in the form $a(x + b)^2 + c$ **(3)**
b) Sketch the curve with equation $y = 4x^2 - 16x + 19$ **(2)**
c) Find the equation of the straight line which passes through the points P and T. **(3)**

(Total 8 marks)

17 $ABCD$ is a square.
A is the point $(0, 5)$ and C is the point $(8, -7)$.
Find the equation of the diagonal BD.
Give your answer in the form $ax + by = c$

(Total 6 marks)

18 *PQRS* is a rhombus.

P has coordinates (0, 10) and *R* has coordinates (12, 6).

Find the equation for the diagonal *QS*, giving your answer in the form $y = mx + c$

(Total 5 marks)

19 A bag contains 2p coins, 5p coins and 10p coins only.

The ratio of the number of 2p coins to the number of 5p coins to the number of 10p coins in the bag is 2 : 4 : 7

The total value of the 10p coins is £10 more than the total value of the 5p coins.

Work out the total value of the coins in the bag.

(Total 4 marks)

20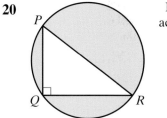

Diagram **NOT** accurately drawn

The diagram shows a right-angled triangle *PQR* and a circle.

P, *Q* and *R* are points on the circle.

$PQ = 12$ cm, $PR = 13$ cm.

Work out the area of the shaded part of the circle.

Give your answer correct to 3 significant figures.

(Total 5 marks)

21 The diagram shows two circles.

A, *B*, *D* and *E* are points on one circle.

B, *C* and *D* are points on the second circle.

BC is a diameter of the second circle.

ABC and *CDE* are straight lines.

a) Show that angle $BAE = 90°$

You must give a reason for each stage in your working. **(4)**

$AB = 7$cm, $BC = 5$ cm, $DE = 10$ cm and $CD = x$ cm

b) i) Show that $x^2 + 10x - 60 = 0$

ii) Calculate the length of *CE*.

Give your answer correct to 3 significant figures. **(6)**

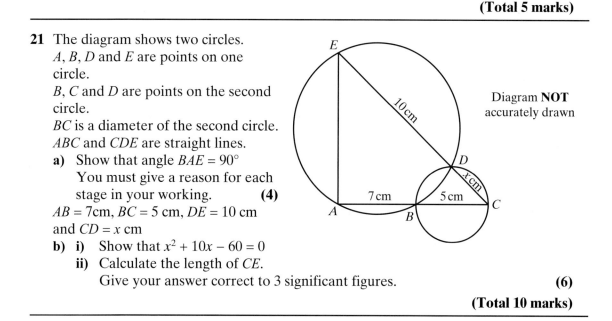

Diagram **NOT** accurately drawn

(Total 10 marks)

22 A coin is biased.

When it is spun, the probability that it shows Heads is p.

Shauna spins the coin twice.

The probability that the coin shows Tails both times is 4 times the probability that it shows Heads both times.

  a) Show that $3p^2 + 2p - 1 = 0$ **(2)**

  b) **i)** Solve $3x^2 + 2x - 1 = 0$

        ii) State the value of p.   **(4)**

                                                    (Total 6 marks)

23

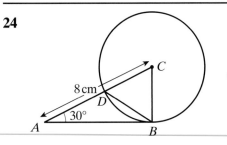

Diagram **NOT** accurately drawn

The diagram shows the lengths, in centimetres, of sides AB and AC of triangle ABC.

Angle $BAC = 30°$

The area of triangle ABC is 24 cm^2

  a) Show that $x^2 - 2x - 99 = 0$ **(3)**

  b) Find the length of AB.   **(3)**

                                                      (Total 6 marks)

24

Diagram **NOT** accurately drawn

B and D are points on a circle, centre C.

AB is a tangent to the circle.

ADC is a straight line.

$AC = 8$ cm.

Calculate the length of BD.

                                                      (Total 4 marks)

25 In triangle ABC, $AB = 10$ cm and $AC = 9$ cm.

Angle A is obtuse.

The area of triangle ABC is 40 cm^2.

Calculate the length of BC.

Give your answer correct to 3 significant figures.

Diagram **NOT** accurately drawn

                                                      (Total 6 marks)

26

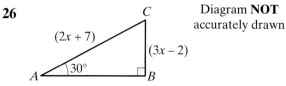

Diagram **NOT** accurately drawn

The diagram shows the lengths, in centimetres, of two sides of triangle *ABC*.
Angle *ABC* = 90°
Angle *BAC* = 30°
Find the value of *x*.

(Total 5 marks)

27

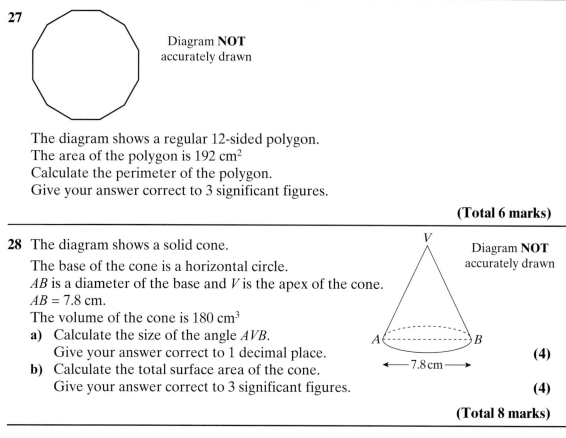

Diagram **NOT** accurately drawn

The diagram shows a regular 12-sided polygon.
The area of the polygon is 192 cm²
Calculate the perimeter of the polygon.
Give your answer correct to 3 significant figures.

(Total 6 marks)

28 The diagram shows a solid cone.

The base of the cone is a horizontal circle.
AB is a diameter of the base and *V* is the apex of the cone.
AB = 7.8 cm.
The volume of the cone is 180 cm³

a) Calculate the size of the angle *AVB*.
Give your answer correct to 1 decimal place. **(4)**

b) Calculate the total surface area of the cone.
Give your answer correct to 3 significant figures. **(4)**

Diagram **NOT** accurately drawn

(Total 8 marks)

29 The parabola with equation $y = x^2 + ax + b$ has a minimum point at $(2, -3)$.
Find the value of *a* and the value of *b*.

(Total 3 marks)

30 Shape **P** is mapped onto shape **Q** by a reflection in the x-axis.

Shape **Q** is mapped onto shape **R** by a reflection in the y-axis.
a) Describe fully the single transformation which maps shape **P** onto shape **R**. **(2)**
The graph of $y = x^2 + 4x - 5$ is rotated through $180°$ about the origin.
b) Find the equation of the new graph. **(2)**

(Total 4 marks)

---

31 The graph of $y = 3x - 2$ is reflected in the line with equation $y = x$

Find the equation of the new graph.

(Total 2 marks)

---

32 The table shows information about the number of films watched on TV by a group of students last month.

The first frequency is obscured by a blot of ink.

Number of films	Frequency	Midpoint
0 – 6	🔲	3
7 – 13	10	10
14 – 20	22	17
21 – 27	12	24

Using the midpoints, a correct estimate for the mean number of films watched is 15.6
Find the total number of students in the group.

(Total 5 marks)

---

33 The diagram shows the length, in centimetres, of each side of triangle ABC.

Angle $BAC = 60°$
Calculate the value of x.

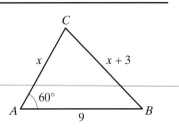

Diagram **NOT** accurately drawn

(Total 4 marks)

---

34 The diagram shows the length of each side of triangle ABC.
Angle $BAC = 60°$
a) Find the two possible values of x. **(4)**
b) Calculate the greatest possible size of angle B.
Give your answer correct to 1 decimal place. **(3)**

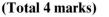

Diagram **NOT** accurately drawn

(Total 7 marks)

35

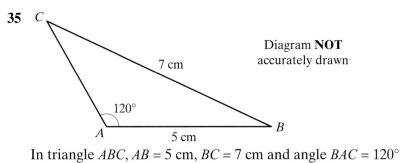

Diagram **NOT** accurately drawn

In triangle *ABC*, *AB* = 5 cm, *BC* = 7 cm and angle *BAC* = 120°
Calculate the length of *AC*.

(Total 4 marks)

36

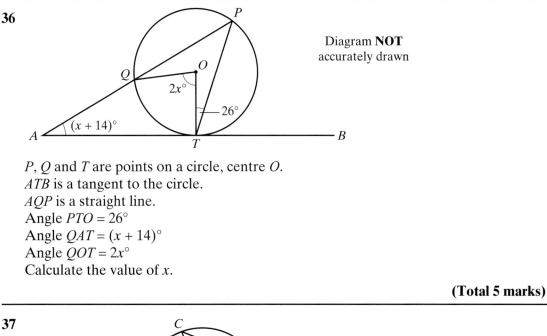

Diagram **NOT** accurately drawn

P, *Q* and *T* are points on a circle, centre *O*.
ATB is a tangent to the circle.
AQP is a straight line.
Angle *PTO* = 26°
Angle *QAT* = (*x* + 14)°
Angle *QOT* = 2*x*°
Calculate the value of *x*.

(Total 5 marks)

37

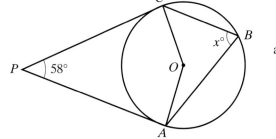

Diagram **NOT** accurately drawn

A, *B* and *C* are points on a circle, centre *O*.
Angle *ABC* = *x*°
PA and *PC* are tangents to the circle.
Find the value of *x*.

(Total 5 marks)

Answers

NUMBER

1 Decimals

1. £46.74
2. a) £14.40 b) i) £56.00
   ii) 3 hours 53 min 20 sec
3. a) £55.63 b) £830
4. £35.36
   1314
   £123.91
   £159.27
   £7.96
   £167.23
5. 4.4
6. 3.5
7. 8.2352941…
8. a) 1.4541019… b) 1.45
9. a) 0.0055555… b) 0.0056
10. a) 8.50121577… b) 8.50
13. $\frac{7}{11}$
16. $\frac{29}{45}$
17. a) for example 63
    b) multiple of 21

2 Powers and roots

1. 5.74569404…
2. a) 1.074172311… b) 1.07
3. a) 17.08016258 b) 17.08
4. a) 5.329668024… b) 5.330
5. $5 \times 5 \times 7$
6. $2^3 \times 7^2$
7. $2^2 \times 3 \times 5^2$
8. a) i) $2^2 \times 3^3$ ii) $2^4 \times 3^2$
   b) 36
9. a) $2 \times 3 \times 5 \times 5$ b) 450
10. a) 21 b) 735
11. a) $(6 + 4) \times 2 = 20$
    b) $6 \div 4 + 2 = 3.5$
    c) $6 \div (4 \times 2) = \frac{3}{4}$
12. a) $\frac{1}{3}$ b) $0.\dot{3}$
13. a) i) 5^{12} ii) 9^3 b) $n = 9$
14. a) i) 4^8 ii) 6^{-2} b) 0
15. a) $\frac{1}{81}$ b) $\frac{4}{9}$ c) $\frac{9}{64}$
16. a) $3^{\frac{3}{2}}$ b) $8^{\frac{1}{6}}$ c) $2^{-\frac{7}{2}}$
17. $x = 3, y = 13$
18. $x = 5, y = 14$
19. $k = 5, n = 3$
20. $n = 10$
22. $7\sqrt{3}$
24. $13 + 4\sqrt{3}$
25. $a = 2, b = 8$
26. $2\sqrt{3}$
27. $\dfrac{\sqrt{15}}{5}$
28. $\dfrac{\sqrt{2}}{16}$
29. a) $4 - 2\sqrt{3}$
30. $\dfrac{3 - \sqrt{5}}{4}$
31. $3 + 2\sqrt{3}$
32. $5 + 4\sqrt{2}$
33. $-8 + 3\sqrt{7}$

3 Fractions

17. $\frac{5}{8}$
18. 12
20. 25
21. $\frac{1}{3}$
22. 14 850 m^3
23. $\frac{2}{3}$
24. 24
25. a) $3000 b) $\frac{2}{3}$

4 Percentages

1. 18%
2. 14%
3. 4%
4. $7905
5. a) £39.60 b) £575
6. £850
7. 175
8. 32%
9. a) 72% b) 140
10. £533
11. 16%
12. £120
13. 6.4%
14. £29 260
15. 12%
16. £180 000
17. 510 000 000 km^2
18. a) 12% b) 115% c) 24 g
19. 9952
20. £1020
21. $156
22. 4.5% per annum
23. 81.6 cm
24. a) £198 b) £520
25. £6615
26. $624.32
27. £8201.25
28. 6
29. £27 500
30. 14
31. 82%
32. 32%
33. No (it's the same as an
    increase of 4%)
34. 17%
35. Decrease of 28%

5 Ratio and proportion

1. 1 : 0.08
2. 1 : 150
3. a) 1:25 b) 9 m
4. 60 300
5. $336
6. a) 3.25 kg copper,
   1.75 kg zinc b) 650 grams
7. 24.2 cm

---

8 105°

9 $18\frac{2}{3}$

10 a) 39 **b)** \$1134

11 a) 180 km **b)** 6 mm

12 375 g of pastry, 1125 g of potatoes, 550 g of onions, 800 g of bacon

13 a) £1242 **b)** €300

6 Standard form

1 a) 2.3×10^5 **b)** 0.0063

2 a) 4.7×10^{-4} **b)** 3 700 000

 c) 6.25×10^7

3 a) 5.8×10^7

 b) 2.9×10^9 km

4 4.8×10^2

5 7.67×10^4

6 a) 1.5×10^{10}

 b) 15 gigayears

7 a) 1.67×10^{-24}

 b) 9.0×10^{-28} g

8 1.42×10^8

9 5.8×10^8 km²

10 5.13×10^{-5}

11 a) 1.23×10^{13} **b)** 1.8×10^{17}

 c) 8.1×10^{-5} nanoseconds

12 2.8×10^{-5}

13 2.3×10^6 years

14 7.67×10^{-4}

15 8.3×10^{-3} cm

16 7.1×10^8

17 a) 1.992×10^{-26} kg **b)** 197

18 a) 4.1×10^{13} km

 b) 2.8×10^4 light years

19 a) 0.000 002 2

 b) 2.91×10^{-13} s

20 a) 81 **b)** 5.8×10^{20}

21 $6 \times 10^{m+n}$

22 i) $2.4 \times 10^{m+n+1}$

 ii) $1.5 \times 10^{m-n}$

23 $4.9 \times 10^{2n+1}$

24 $6.25 \times 10^{m-n-1}$

25 6

26 $a = 9$ $n = -10$

27 $a = 8$ $n = -4$

28 a) 5 **b)** −3

7 Degree of accuracy

1 3, 30, 0.4

2 a) 40, 9, 30 **b)** 12

 c) Both numbers in the numerator have been rounded down and the number in the denominator has been rounded up.

3 a) 40, 80, 0.5, 4 **b)** 1600

 c) Both numbers in the numerator have been rounded up; both numbers in the denominator have been rounded down (and the final rounding was up).

4 i) 6.45 kg **ii)** 6.35 kg

5 i) 11.85 cm **ii)** 11.75 cm

6 22.6 kg

7 a) i) 29.8 cm **ii)** 29.4 cm

 b) i) 30 cm **ii)** Upper and lower bounds agree to 1 significant figure.

8 1077 km

9 a) 2 **b)** 61.75

10 4.7 cm

11 0.13 m or 13 cm

12 3

13 a) 7.464366831

 b) Lower bound = $\dfrac{23.35}{\pi}$

 = 7.432535842 and so upper and lower bounds agree to 1 sf i.e. 7 cm

14 6.210526316 cm

15 842.625 km

16 7.025641026 m/s

17 35

18 14 cm

19 23 (Upper bound = 23.06805075 lower bound = 22.50137724 so bounds agree to 2 sf)

20 0.8 (Upper bound = 0.793103448 lower bound = 0.768707483 so bounds agree to 1 sf)

8 Set language and notation

1 a) i) 1, 2, 3, 4, 5, 6, 8 **ii)** 1, 3

 b) 5 is not a member of set A.

2 a) i) rhombuses, squares, kites

 ii) squares

 b) i) $21 \notin \mathscr{E}$

 ii) No, $14 \in P$ and $14 \in Q$

3 a) 3, 4, 6, 8 **b)** eg 1, 2, 5, 7

4 a) i) {9, 10, 11, 12, 13}

 ii) {14, 15} **b)** ∅

5 a) 21, 23, 28, 29 **b)** ∅ **c)** No; 27 is not a prime number

6 a) i) $12 \notin \mathscr{E}$ since 12 is not odd.

 ii) 1, 3, 5, 15

 b) 1, 3

7

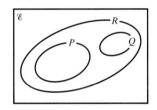

8

 a) $A \cap B' = A$

 b) $A' \cup B = \mathscr{E}$

9 a) i)

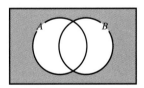

 ii) eg 3, 7 **iii)** eg 5, 15

 b) i)

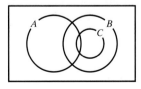

 ii) 3

10 a) i) 5 **ii)** 15 **iii)** 6

 b) i) 25 **ii)** 15 **iii)** 0

11 **a) i)** $13 - x$ **ii)** $10 - x$
  b) 8
12 15

ALGEBRA

9 Algebraic manipulation

1 **a)** $6b - 5a - 1$ **b)** $6p + 21$
 c) $3(3x + 4)$
2 **a)** $3x^2 + 2x$ **b)** $15x + 11$
 c) $y^2 + 6y + 8$
3 **a)** $8x + 11$ **b)** $5x^2 + 4xy$
 c) $4(3y - 2)$
4 **a)** $9y - 13$ **b)** $5(3 - 4p)$
 c) $x^2 - 5x - 6$
5 **a)** $7q^3 + 5q$ **b)** $-3y - 7$
 c) $x(x - 6)$
6 **a)** $11 - 5y$ **b)** $2x^2 - 6xy$
7 **a)** $14x - 21$ **b)** $3p^4 - 3p^3$
 c) $y^2 - 6y + 5$
8 $x^2y^2(x^2 - y)$
9 $5ef^2(4e^2 + 5)$
10 $9c^2d(3d^3 - 2c)$
11 **a)** $2x(9 - 4x)$
 b) $(x + 4)(x + 5)$
12 **a)** $6p^2 + 11pq - 10q^2$
 b) $(x + 4)(x - 4)$
13 $(x + 2)(x - 10)$
14 $(x + 5)(2x - 1)$
15 $2(4x - 3)(x + 2)$
16 $3(2x + 5)(2x - 5)$
17 $\frac{x+2}{4}$
18 $\frac{x+3}{x+1}$
19 $\frac{x-3}{x-12}$
20 $\frac{x+2}{x-2}$
21 $\frac{x+4}{2x-3}$
22 $\frac{5}{x-3}$
23 $\frac{2}{x+4}$
24 **a)** $(3x + 1)(3x - 1)$
 b) i) 29×31
 ii) $2 \times 2 \times 5 \times 5 \times 29 \times 31$

25 $\frac{2}{x-6}$
26 $\frac{2x}{x-3}$
27 $x^3 - 4x$
28 $x^3 + 6x^2 + 11x + 6$
29 $x^3 - 6x^2 + 11x - 6$
30 $2x^3 - 7x^2 - 5x + 4$
31 $-x^3 + 2x^2 + 9x - 18$
32 $x^3 - 4x^2 - x + 4$
33 $3x^3 - 8x^2 + 7x - 2$
34 **a)** $x^3 + 3x^2 - x - 3$
 b) $2x^4 + 6x^3 - 2x^2 - 6x$
35 **a)** $(x-3)^2 + 4$
 b) i) 4 **ii)** 3
36 **a)** $2(x + 1)^2 + 5$ **b)** $(-1,5)$
37 **a)** $7 - (x+2)^2$ **b)** 7
 c) $x = -2$
38 **a)** $a = 2, b = 5$
39 **a)** $(x-4)^2 + 2$
40 **a)** $3(x-1)^2 - 2$
 b) $-2 < k < 1$

10 Expressions and formulae

1 27
2 6
3 4.4
4 4.6
5 $\frac{8}{15}$
6 $F = 40H + 32$
7 **a)** 50 **b)** $P = 2(2n+1)$
 c) $n = \frac{P-2}{4}$

8 Either $r = +\sqrt{\dfrac{A}{\pi}}$ or
 $r = -\sqrt{\dfrac{A}{\pi}}$
9 $h = \dfrac{mx - y}{m}$
10 $s = \dfrac{at^2}{2}$
11 $x = \dfrac{2y - 6b}{3 + 5y}$ or $x = \dfrac{2(y - 3b)}{3 + 5y}$
12 $x = \dfrac{a}{y} - a$
13 Either $r = +\sqrt{\dfrac{aR^2 + t - T}{a}}$
 or $r = -\sqrt{\dfrac{aR^2 + t - T}{a}}$
14 **a)** 44 **b)** $P = 4\sqrt{A}$
 c) $A + \dfrac{P^2}{16}$
15 $t = \sqrt{\dfrac{2(3s+13)}{a}}$

11 Linear equations and simultaneous equations

1 $x = \frac{8}{3}$
2 $x = -\frac{5}{6}$
3 $y = \frac{1}{2}$
4 $x = 12$
5 $x = \frac{3}{2}$
6 $x = \frac{1}{2}$
7 $x = -\frac{3}{8}$
8 $x = \frac{1}{2}$
9 $x = \frac{9}{5}$
10 $x = \frac{1}{10}$
11 $y = \frac{11}{5}$
12 $p = -\frac{4}{3}$
13 **a)** $6x - 1 = 14$ **b)** $x = \frac{5}{2}$
 c) 6 cm

14 **a)** $4x + 10 = 2(x + 10)$
b) 5 years
15 **a)** $3x - 8 = 2(x + 8)$ **b)** 84°
16 $x = -\frac{1}{2}$
17 $t = 1.4$
18 $a = 1.6$
19 $x = -0.5, y = -4.5$
20 $x = 1.5, y = -1$
21 $x = 3, y = 1.5$
22 **a)** $x = \frac{3}{4}, y = -\frac{1}{2}$ **b)** $(\frac{3}{4}, -\frac{1}{2})$
23 $(\frac{5}{4}, \frac{1}{2})$

12 Coordinates and graphs

1 $(3, 7)$
2 $(1, 8)$
3 **a)** $(-1, 8)$ **b)** $(1.5, 6.5)$
4 **a)** 1230 **b)** 10 **c)** 100
d) 1210 and 1230
e) 1225, 1250 and 1305
f) 80 km/h
5 **a)** 10 **b)** 12 km
c) 12 km/h **d) i)**

ii) 105
6 **a)** 2.8 m/s² **b) i)** 28 m/s
ii) 504 m
c)

7 **a)** 25 m **b)** 45 m
c) 2 seconds

13 Linear graphs

1 **b)** $(-1, 4)$
2 **b)** $(-2, -3)$
4 **b)** eg $x + y - 3 = 0$.
5 $\frac{2}{7}$
6 **a)** 2 **b)** $y = 2x - 2$
7 **a) i)** 0.25 **ii)** Cost, in pounds, of a 1 minute call
b) $y = 0.25x + 12$ **c)** 80
8 **a)** $y = 8x + 9$ **b)** $(0, -2)$
9 $y = -1.5x + 7$
10 **a)** -2 **b)** $y = -2x + 1$
c) -5
11 3
12 0.5
13 1.5
15 $3x + 4y = 20$
16 -5
17 $y = 2x - 5$
18 $y = 4x - 1$

14 Sequences

1 **a)** 35, 48 **b)** 98
2 $6n - 1$
3 $46 - 4n$
4 **a)** 22 **b)** 39 **c)** $119 - 10n$
d) 17
5 46
6 **a)** $8n - 10$ **b)** 92
7 **a)** $5n + 1$ **b)** 101
c) 1070
8 **a)** 3 **b)** 3060
9 **a)** 29 **b)** 50
10 9024
11 1215
12 -5
13 **a)** -4 **b)** 63
14 **a)** -12 **b)** 400
15 **a)** $3a + 3d$
16 **a)** 4950
b) 1647, 1650, 1653
17 -8050
19 165°
20 **a)** 1665 **b)** 4905
c) 3240

15 Quadratic equations

2 $x = 0, x = 4$
3 $y = -9, y = 12$
4 $x = -\frac{1}{2}, x = 3$
5 $x = 1, x = 8$
6 $x = -5.56, x = -1.44$
7 $y = -0.717, y = 1.12$
8 $p = -6.46, p = 0.46$
9 $x = -\frac{3}{2}, y = \frac{9}{2}$ and $x = 3$, $y = 18$
10 $x = -\frac{2}{5}, y = 2\frac{1}{5}$ and $x = -2$, $y = -1$
11 $x = -1\frac{1}{5}, y = 3\frac{2}{5}$ and $x = 2$, $y = -3$
12 $x = -2, y = -1$ and $x = 4$, $y = 2$
13 **a)** $(x + 2)^2 = \frac{1}{2}(2x + 1)^2$
c) 12.5 cm
14 **a)** $(x - 2)^2 - 3$
b) $2 \pm \sqrt{3}$

16 Inequalities

1 $x > 3$
2 $x < 2.5$
3 $-2 \leqslant x < 3$
4 $-1 < x \leqslant 3$
5 $-1.5 < x < 2$
6 **a)** $x > -3$
b)

7 **a)** $x \leqslant 3$
b)

8 **a)** $-2 \leqslant x < 4$
b)

9 **a)** $x < 5$ **b)** 1, 2, 3, 4
10 **a)** $x > -3.5$ **b)** $-3, -2, -1$
11 **a)** $-2 \leqslant x < 1.5$
b) $-2, -1, 0, 1$

12

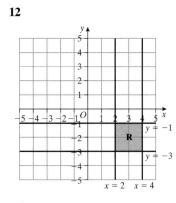

$y = -1$
$y = -3$
R
$x = 2$ $x = 4$

13

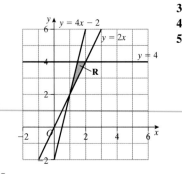

R
$y = 2$
$x + y = 3$

14

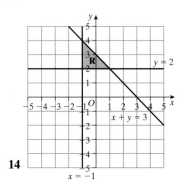

$y = 4x - 2$
$y = 2x$
$y = 4$
R
$x = -1$

15

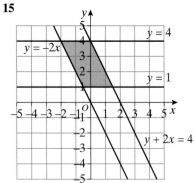

$y = 4$
$y = -2x$
$y = 1$
$y + 2x = 4$

16 a) $-3 < x < 3$

b)

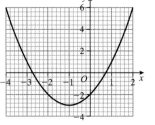

17 a) $x \leqslant -2$ or $x \geqslant 2$

b)

18 a) $-3 \leqslant x \leqslant 2$

b)

19 a) $x > 3$, $x < -1$

b)

20 $0 \leqslant x \leqslant 2$

21 a) $x < -\frac{1}{2}$, $x > 2$ **b)** 3

22 $\frac{2}{3} < x < \frac{5}{2}$ **b)** 1, 2

17 Indices

1   **a)** a^7 **b)** $4p^6$ **c)** q^4
2   **a)** 16   **b) i)** c^9 **ii)** d^5
3   **a)** -6 **b)** t^4
4   **a)** 20   **b)** $2p^7q^5$ **c)** $48r^2$ **d)** 5
5   **a)**

x	-4	-3	-2	-1
y	6	1	(-2)	-3

x	0	1	2
y	(-2)	1	(6)

b)

6   **a)** x^3 **b)** y^2
7   **a)** $27a^{12}$ **b)** $2p^3$
8   **a)** 1   **b)** $16p^{12}q^{16}$ **c)** $r^{-6}t^8$
9   **a)** $16x^8$ **b)** $\frac{2}{x}$

18 Proportion

1   **a)** $E = 0.25T$ **b)** 2.5
c) 26 newtons
2   **a) i)** $A = 0.5D^2$ **b)** 84.5
c) 18 cm
3   **a)** $y = \frac{3\sqrt{x}}{2}$ **b)** 3.75
c) 4096
4   **a)** $M = 0.016r^3$ **b)** 6.75 kg
c) 50 cm
5   **a) i)** $V = \frac{1000}{p}$

ii)

V
O
P

b) 2 m³
c) 125 newtons/m²

6   **a)** $d = \frac{2.4}{\sqrt{h}}$ **b)** 0.75   **c)** 0.64

7   **a)** $F = \frac{32}{D^2}$ **b)** 1.28   **c)** 1.6

8   **a)** $x = 4.9t^2$ **b)** 122.5
c) $1\frac{1}{7}$ seconds

9   **a)** $T = \frac{40}{\sqrt{h}}$ **b)** 5   **c)** 0.25

10   **a)** $T = \sqrt{\frac{9.8}{g}}$ **b)** 2.47 s
c) 3.69 m/s²

11   **a)** $C = \frac{135}{v^3}$ **b)** 3.75

12   **a)** $N = \frac{4000}{r^3}$ **b)** 32

19 Function notation and transformation of functions

1 **a)** 4 **b)** 1.5
2 **a)** $1 - 3x$ **b)** $5 - 3x$
3 $f^{-1} : x \mapsto \dfrac{3}{x} - 2, x \neq 0$
4 $\dfrac{1}{4}$
5 **a)** -9 **b)** $-1.6, 0.2, 3$ **c)** 3
6 **a)** -9 **b)** 1.5 **c)** 6
  d) $x < -1$
  e) $g^{-1} : x \mapsto x^2 - 1$
7 **a) i)** 2 **ii)** -7 **b)** 1
  c) i) $ff(x) = x$ **ii)** f^{-1} is the
  same as f
8 $x = 1, x = 1\dfrac{7}{8}$
9 **a) i)** -3 **ii)** -2
  b) $f^{-1} : x \mapsto \dfrac{x + 3}{4}$
  c) i) $gf : x \mapsto \dfrac{4}{4x - 3}$ **ii)** 0.75
10 **a)** $\dfrac{4x - 2}{3}$ **b)** $x = -6$
11 **a)** $y = x$ **b)** $x = -5$
12 **a)** $(x + 1)^2 + 2$
  b) $f(x) \geqslant 3$
  c) $f^{-1} : x \mapsto \sqrt{x - 2} - 1$
13 **a)** $2\left(x - \dfrac{5}{2}\right)^2 - \dfrac{35}{2}$
  b) $f(x) \geqslant -\dfrac{35}{2}$
14 $g^{-1} : x \mapsto \sqrt{\dfrac{x - 7}{3}} - 1$
15 translation with vector $\begin{pmatrix} 0 \\ 5 \end{pmatrix}$
16 $y = -x^2 + 3x - 1$
17 $y = x^2 + 5x + 3$
18 $y = \sin(x - 30)°$
19 translation with vector $\begin{pmatrix} -60 \\ 2 \end{pmatrix}$
20 reflection in the y-axis or a
  translation with vector $\begin{pmatrix} -2 \\ 0 \end{pmatrix}$

21 **a) i)** $(2, 3)$ **ii)** $(2, 3)$
  iii) $(2, -15)$ **iv)** $(6, -3)$
  v) $(1, -3)$
  b) 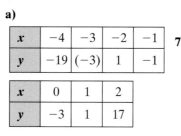 $\begin{pmatrix} -2 \\ 3 \end{pmatrix}$

20 Harder graphs

1 **a)**

x	-4	-3	-2	-1
y	-19	(-3)	1	-1

x	0	1	2
y	-3	1	17

b)
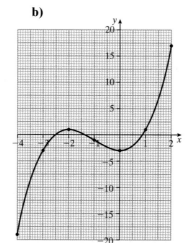

c) i) $-2.6, -1.3, 0.9$ **ii)** $-2, 1$
2 **a)** $(-1, 7)$ **b)** $1.6, -3.6$
  c) $-3.8, 0.8$
3 **a)**

x	0.4	0.6	0.8	1
y	10.4	(7.3)	5.8	(5)

x	1.5	2	3	4
y	(4.2)	4	(4.3)	5

x	5	6
y	(5.8)	(6.7)

c) $0.75, 5.2$ **d)** $y = 10 - 2x$

4 **a)** -3 **b)** $-1.9, 0.3, 1.5$
5 124 (accept answers in range 115 to 135).
6

x	1	2	3	4
y	500	150	93.3	80

x	5	6	7
y	79.2	83.3	89.9

7

Equation	Graph
$y = \sin x$	C
$y = \cos x$	A
$y = \tan x$	B

8 **a)**

x	y
0	0
30	1.5
60	2.6
90	3
120	2.6
150	1.5
180	0
210	-1.5
240	-2.6
270	-3
300	-2.6
330	-1.5
360	0

9
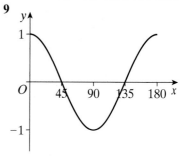

21 Calculus

1 **a)** $8x - 4$ **b)** 0

2 $\dfrac{dy}{dx} = 8x - \dfrac{1}{x^2}; (\frac{1}{2}, 3)$

3 $\frac{1}{3}$ m/s^2

4 **a)** -2 **b)** $\dfrac{2x}{3} - 4$

5 **a)** $3t^2 + 6t - 9$ **b)** 12 m/s^2

6 **a) i)** $6 - 4x$ **ii)** $(1.5, 11.5)$
   b) maximum; eg coefficient
   of x^2 is negative **c)** $x = 1.5$

7 **a) i)** $3x^2 - 6x - 9$
   ii) $x_A = -1$, $x_B = 3$
   b) 13 and -19

8 **b)** $840 - 60x$ **c)** $780

9 **a)** $3x^2 + 75$ **b)** $(2, 58)$ and
   $(-2, -258)$ **c)** none

10 **b)** $8x - \dfrac{216}{x^2}$ **c)** 4 cm

11 **a)** $-2, 0$ and 2

   b) $(-2, 64)$ and $(2, -64)$

SHAPE, SPACE AND MEASURES

22 Compound measures

1 24 km/h
2 76 km/h
3 11.63 s
4 2 hours 24 minutes
5 19.8 km
6 712 km/h
7 3567 km
8 12 hours 25 minutes
9 0.006 m/s
10 16 hours 36 minutes

11 **i)** 7.86 g/cm^3 **ii)** 7860 kg/m^3

12 1000 grams

13 0.0149 m^3
14 1500 Pascals
15 2540 N
16 0.02 m^2
17 5 604 000
18 45 l
19 18.2 l
20 429 000 J

23 Construction

5 **a)** 2.1 m **b)** 64°

6 420 m (accept answers in
  range 415 to 430)

7 246 m (accept 240 m to
  250 m)

8 **a)** 220 km **b)** 280°

24 Geometry

1 **a)** $x = 76$ [(vertically)
   opposite angles;
   corresponding angles]
   b) $y = 67$ [sum of angles
   on a straight line = 180°;
   alternate angles (or
   corresponding angles)]

2 $\angle PTQ = 69°$ [$\angle PQT = 63°$
   (corresponding angles)
   $\angle PTQ = 180° - (63° + 48°)$
   (angle sum of triangle)]

3 43° [alternate angles;
   (vertically) opposite angles;
   angle sum of triangle]

4 116° [alternate angles;
   base angles of an isosceles
   triangle; angle sum of a
   triangle]

5 **a)** 1080° **b)** 36°
6 **a)** 140° **b)** 40°
7 **a)** 12 **b)** 64°
8 **a)** 18 **b)** 2880°
9 15
10 20
11 **a)** 038° **b)** 218°
12 092°

25 Transformations

1 **a)** Rotation of 90° clockwise
   about the origin
   b)

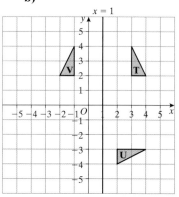

2 **a)** Reflection in the line $y = x$
   b)

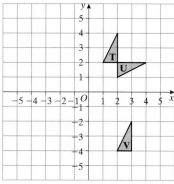

3 **a) b)**

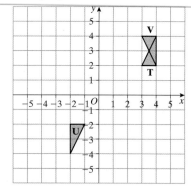

4 a) b)

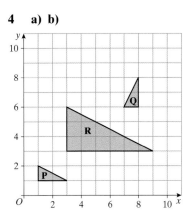

5 **a)** Translation with

vector $\begin{pmatrix} 4 \\ -2 \end{pmatrix}$

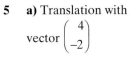

b) i) Enlargement with scale factor 2, centre (2, 0)
ii) Enlargement with scale factor $\frac{1}{2}$, centre (2,0)

6

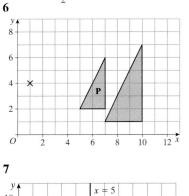

7

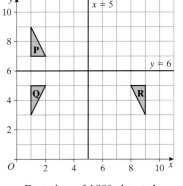

Rotation of 180° about the point (5, 6)

8

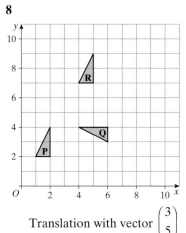

Translation with vector $\begin{pmatrix} 3 \\ 5 \end{pmatrix}$

26 Circle properties

1 29
2 **a)** 33° **b)** A line drawn from the centre of a circle to the midpoint of a chord is perpendicular to the chord. Angle sum of triangle = 180°.
3 28°
4 134°
5 **a) i)** 38° **ii)** Angles in the same segment are equal
   b) 52°
6 **a) i)** 61° **ii)** Sum of opposite angles of a cyclic quadrilateral is 180°
   b) 29°
7 **a) i)** 148° **ii)** Angle at the centre is twice the angle at the circumference
   b) i) 74° **ii)** Angles in the same segment are equal.
8 **a) i)** 39° **ii)** Angle at the centre is twice the angle at the circumference
   b) 31° (Sum of opposite angles of a cyclic quadrilateral is 180°. Angle sum of triangle = 180°. In an isosceles triangle, the angles opposite the equal sides are equal.)

9 **a) i)** 62° **ii)** Alternate segment theorem
   b) 118°
   c) Alternate angles *PTC* and *BCT* are equal
10 **a) i)** 51° **ii)** Alternate segment theorem
   b) 137°
11 8 cm
12 10

27 Area and perimeter

1 51 cm²
2 8
3 110 cm²
4 172 cm²
5 80 cm²
6 26.1 cm
7 88.2 cm²
8 3.77 cm
9 54.5 cm²
10 41.1 cm
11 23.2 cm
12 281 cm²
13 20.1 cm²
14 27.2 cm
15 37 000 cm²

28 3-D Shapes – volume and surface area

1 254 cm²
2 **a)** 3600 cm³ **b)** 2040 cm²
3 **a)** 480 cm³ **b)** 440 cm²
4 **a)** 5090 cm³ **b)** 1640 cm²
5 273 cm²
6 **a)** 437 cm² **b)** 860 cm³
7 208 cm²
8 231 cm³
9 2470 cm³
10 **a)** 82.8 cm³ **b)** 127 cm²
11 37.2 cm²
12 149 cm²
13 214 cm²
14 1.36 cm
15 640 000 cm³

Answers

29 Pythagoras' theorem

1. 7.44 cm
2. 2.4
3. 6.69
4. 8.4 cm
5. 9.88 cm
6. 22.2 km
7. 6.09 cm
8. 32.9 cm^2
9. 17.7 cm^2
10. 8.06 cm
11. 13
12. 49.3 cm^2
13. a) 90°   b) 9.72 cm
14. 7.71 cm
15. 10.6 cm
16. 23.2 cm
17. 132 cm^3
18. 171 cm^2
19. 79.8 cm^3
20. 32.1 cm^2
21. 27.6 cm
22. 8.51 cm
23. 11.5
24. 7.93 cm
25. 44.4 cm^2

30 Trigonometry

1. 3.47 cm
2. 36.7
3. 3.64 cm
4. 61.6°
5. 8.46
6. 2.51 cm
7. 31.9
8. 8.28 cm
9. 1.71 m
10. 076°
11. 293 m
12. 19.8 m
13. 42.7°
14. 10.3 cm
15. 21.9 cm
16. 3.2°
17. 10.7 cm
18. 6.27 cm
19. 12.5°
20. 3.71 cm
21. 43.3
22. 7.99 cm
23. 37.9°
24. 7.55 cm
25. 6.96 cm

31 Similar shapes

1. a) 76°   b) i) 5.4 cm
   ii) 1.2 cm
2. a) 5.4 cm   b) 13.5 cm^2
3. a) 9.1 cm   b) 3.6 cm
4. a) 3.9 cm   b) 5.4 cm^2
5. a) 25 : 16   b) 16 : 9
6. a) 7 cm   b) 6.6 cm
7. a) 3.2 cm   b) 4.2 cm
   c) 16 : 9
8. No; e.g. $\frac{110}{100} = 1.1$
   but $\frac{90}{80} = 1.125$
9. a) 10.2 cm   b) 4.9 cm
   c) 270 cm^2
10. a) 8.1 cm   b) 7.6 cm
    c) 48 cm^2
11. a) 352 cm^2 b) 162 cm^3
12. a) 15 cm   b) 1750 cm^3
13. a) 18 cm   b) 999 cm^2
14. a) 9 cm   b) 216 cm^3
15. a) 303 cm^2 b) 4850 cm^2

32 Advanced trigonometry

1. a) 19.2°   b) 9.11 cm
2. 67.3°
3. 69.0 cm^3
4. 736 cm^2
5. 13.9°
6. a) 16.4 cm^2 b) 6.14 cm
7. 7.87 cm
8. 42.3°
9. 58.2°
10. a) 82.8°   b) 22.9 cm^2
11. 8.76 cm^2
12. 118.5°
13. a) 7.68 cm   b) 22.0 cm^2
14. 509 m
15. a) 82.1 km   b) 095°
16. 11.3 cm^2

33 Vectors

1. a) i) $\begin{pmatrix} 3 \\ 5 \end{pmatrix}$ ii) 5.83   b) (10, 11)
2. a) i) $\begin{pmatrix} 2 \\ -1 \end{pmatrix}$ ii) $\begin{pmatrix} 4 \\ -2 \end{pmatrix}$
   iii) AB and CD are parallel. $CD = 2AB$.   b) 4.47
3. a) $\mathbf{b} - \mathbf{a}$ b) $\frac{1}{2}(\mathbf{b} - \mathbf{a})$
   c) $\frac{1}{2}\mathbf{a} + \frac{1}{2}\mathbf{b}$
4. a) i) $4\mathbf{a}$ ii) $4\mathbf{a} - \mathbf{b}$ iii) $\mathbf{a} + \frac{3}{4}\mathbf{b}$
   b) $\frac{3}{4}$
5. a) $\mathbf{a} + \frac{1}{2}\mathbf{b}$ b) i) $\mathbf{b} + 2\mathbf{a}$
   ii) $\overrightarrow{PT} = 2\overrightarrow{PM}$
6. a) i) $\mathbf{a} - \mathbf{b}$ ii) $\frac{3}{2}\mathbf{a} + \frac{1}{2}\mathbf{b}$
   b) i) $\frac{3}{2}\mathbf{a}$ ii) MN is parallel to PQ and $MN = \frac{3}{4}PQ$.
7. a) i) $\mathbf{a} - \mathbf{b}$ ii) $\frac{2}{3}\mathbf{a} + \frac{1}{3}\mathbf{b}$
   b) i) $\mathbf{a} + \frac{1}{2}\mathbf{b}$ ii) $\overrightarrow{PL} = \frac{2}{3}\overrightarrow{PM}$
8. a) $\mathbf{c} = \mathbf{b} - \mathbf{a}$ b) i) $\frac{3}{2}\mathbf{b}$
   ii) XY is parallel to UT and $XY = \frac{3}{2}UT$
9. a) i) $\mathbf{b} - \mathbf{a}$ ii) $\frac{3}{4}\mathbf{a} + \frac{1}{4}\mathbf{b}$
   iii) $\mathbf{a} + \frac{1}{3}\mathbf{b}$ b) $\overrightarrow{PL} = \frac{3}{4}\overrightarrow{PK}$
10. a) i) $\frac{1}{2}\mathbf{b}$ ii) $\frac{1}{2}\mathbf{b}$
    b) Opposite sides are parallel and equal in length i.e. the quadrilateral is a parallelogram.

HANDLING DATA

34 Graphical representation of data

1   **a)** 10   **b)** 46   **d)** approx 56
2   **b)** cumulative frequencies are: 20, 115, 180, 190, 195
    c) i) approx 70   **ii)** approx 45
3   **a)** 54   **b)** 8   **c)** 6
4   **a)** bar width 25, height 2.2; bar width 10, height 1.6
    b) 24; 35
5   **a)** 48   **b)** 216
6   12%
7   **a) i)** 15   **ii)** 33   **b)** 22

35 Statistical measures

1   **a)** 2 years   **b)** 4.2 years
2   **a)** 0   **b)** 1.25
3   60
4   **a) i)** 30   **ii)** 29.75
    b) i) increase   **ii)** 30 > 29.75
5   **a)** 12   **b)** 2, 8, 8
    c) eg 1, 2, 4, 4
6   61%
7   21 years
8   168 cm
9   26
10 **a)** $0 \leqslant d < 2$ **b)** $2 \leqslant d < 4$
    c) 114 mm
11 **a)** $100 < d \leqslant 110$
    b) $\frac{22}{50}$ or $\frac{11}{25}$ **c)** 97.2 km/h
12 **a)** approx 18 hours
    b) approx 4 hours
13 **a)** 45 000
    b) 4, 20, 36, 46, 50, 52
    d) i) approx 43 000
    ii) approx 17 500
14 **a) i)** 20 years   **ii)** 26 years
    b) eg The ages of the players in the cricket team are more spread out.
15 **a)** 9   **b)** eg The girls' marks were higher and less spread out than the boys' marks.

16 **a)** 15, 110, 155, 170, 180
    c) i) approx 36   **ii)** approx 22
    d) approx 24
17 **a)** approx 34   **b)** approx 34
18 **a)** 60–79   **b)** 58 or 57.5
    c) cumulative frequencies 3, 8, 14, 23, 30

36 Probability

1   **a)** $\frac{75}{100}$ or $\frac{3}{4}$
    b) No; if the coin were fair, it would come down heads approximately 50 times.
2   35
3   21
4   18
5   **a)** 36   **b)** 96
6   **a)** 0.55   **b)** 0.35
7   **a)** 0.2   **b)** 0.65   **c)** 12
8   **a)** 0.24   **b)** 0.56   **c)** No; eg probabilities should be multiplied, not added
9   **a)** 0.6   **b)** 0.45
10 **b)** $\frac{2}{15}$ **c)** $\frac{7}{15}$
11 **b)** $\frac{9}{64}$ **c)** $\frac{30}{64}$ or $\frac{15}{32}$
12 **b)** 12
13 **b) i)** $\frac{2}{15}$ **ii)** $\frac{8}{15}$
14 **b)** $\frac{6}{36}$ or $\frac{1}{6}$ **c)** $\frac{14}{36}$ or $\frac{7}{18}$
15 **b)** $\frac{11}{12}$
16 **a)** $\frac{3}{24}$ or $\frac{1}{8}$ **b)** $\frac{11}{24}$
17 **a)** $\frac{6}{56}$ or $\frac{3}{28}$ **b)** $\frac{26}{56}$ or $\frac{13}{28}$
18 **a)** 0.16   **b)** 0.33
19 **a)** $\frac{3}{40}$ **b)** $\frac{17}{40}$
20 **a)** $\frac{7}{20}$ **b)** $\frac{13}{20}$
21 **a)** 0.027 **b)** 0.784
22 **a)** $\frac{3}{190}$ **b)** $\frac{241}{285}$
23 $\frac{6}{216}$ or $\frac{1}{36}$
24 **a)** 0.32   **b)** 0.53
26 **a)** $\frac{2}{5}$ **b)** $\frac{1}{10}$ **c)** $\frac{3}{7}$

37 Further problem solving and mathematical reasoning

1   Falcons 4   Kestrels 3
2   40
3   600
4   124
5   13.6 cm.
6   75.4 cm²
7   **a) ii)** 406 m
    b) $P = 2\pi w + 400$
8   9.5°
10 4.122×10^5
12 **a)** -13
    b)

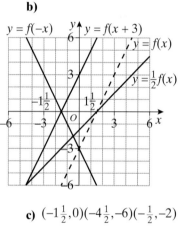

    c) $(-1\frac{1}{2}, 0)(-4\frac{1}{2}, -6)(-\frac{1}{2}, -2)$
13 **a)** $g(x) = -1 - 2x - x^2$
    b) $h(x) = 8 - x^2$
14 **a) ii)** $(-90, -1)\,(90, 1)$
    b) ii) $(-180, -1)\,(0, 1)$ $(180, -1)$
    c) e.g. -90 (in general, $-90 + 360k$, where k is any integer)
16 **a)** $4(x-2)^2 + 3$
    c) $y = -8x + 19$
17 $2x - 3y = 11$
18 $y = 3x - 10$
19 £18.80
20 103 cm²
21 **b) ii)** 14.2 cm
22 **b) i)** $\frac{1}{3}, -1$ **ii)** $\frac{1}{3}$
23 **b)** 12 cm
24 4 cm

25 16.2 cm

26 2.75

27 49.7 cm

28 **a)** 38.1°   **b)** 194 cm^2

29 $a = -4$, $b = 1$

30 **a)** Rotation of 180° about the origin
    b) $y = -x^2 + 4x + 5$

31 $y = \dfrac{x + 2}{3}$

32 50

33 4.8

34 **a)** 3, 5   **b)** 38.2°

35 3 cm

36 25

37 61